1
2
3
4

57
60
58
11
U.S. AIR FORCE
847

MILITARY POWER

RED FLAG

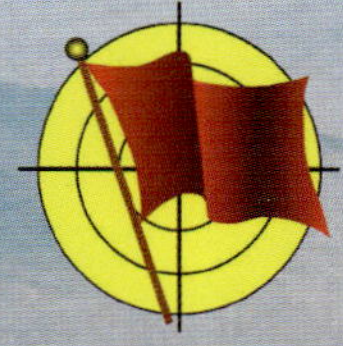

AIR COMBAT FOR THE 21ST CENTURY

TYSON V. RININGER

ZENITH PRESS

In Memory of George Hall, 1941–2006

One of the greatest aviation photographers of all time

My Friend, My Mentor, My Hero

First published in 2006 by Zenith Press, an imprint of MBI Publishing Company, Galtier Plaza, Suite 200, 380 Jackson Street, St Paul, MN, 55101-3885 USA

On the cover: A Blue Team F-16 from the 388th Wing, 421st Fighter Squadron, the Black Widows, flies high above the Nevada desert during a training exercise.

On the frontispiece: Polished brass knobs help make up some of the controls of this KC-135. Serving the military since 1956, these aircraft are among the oldest airframes flying today. This KC-135, and other support crew and aircraft, insure the Red Flag exercises can continue.

On the title pages: F-5E Tiger IIs in wedge formation over Lake Mead, Nevada, in 1981. *USAF*

On the back cover: *(left)*A German MiG-29 flies in formation with an F-15C. *USAF*

*(right)*F-16Cs, flying in left echelon formation, return from the Nellis Range.

About the author: Tyson V. Rininger is the top photographer for the International Council of Air Shows. His photos and articles have been published in numerous magazines including multiple covers for *AirShows* magazine and *World Airshow News*. He also works with many other publications, including *Air & Space Smithsonian*, *Aviation Week & Space Technology*, *Flight Journal*, *Aircraft Illustrated*, and *Combat Aircraft* magazine. Rininger resides in Monterey, California.

Library of Congress Cataloging-in-Publication Data

Rininger, Tyson V., 1974-
Red flag : air combat for the 21st century / Tyson V. Rininger.
p. cm.
Includes index.
ISBN-13: 978-0-7603-2530-8 (softbound)
ISBN-10: 0-7603-2530-8 (softbound)
1. Air warfare. 2. Nellis Air Force Base (Nev.) 3. Airplanes, Military—United States. 4. War games—United States. I. Title.

UG630.R52 2006
358.4'1480973—dc22

2006021734

Editor: Steve Gansen & Scott Pearson
Designer: Brenda C. Canales

Printed in China

Contents

Acknowledgments

Thanks to those who helped me in acquiring information as well as setting up photoshoots:

22nd Air Refueling Wing—Major Bruce McNaughton, Captain Donna Mae Chun, Captain Dean Peterson, Captain Rocky Zacchevs, Second Lieutenant Sterling Boyer, Senior Airman Scott Scurlock, and Senior Airman Rory Wilcox

64th Aggressor Squadron—Major Andy "Popeye" Hansen and Major Derek "Tazz" Route

141st Air Refueling Wing—Lieutenant Colonel Larry Bruce, Lieutenant Colonel Patricia Morales, Lieutenant Colonel Nancy Reid, Major David Kimpel, Captain Molly Marshall, and Master Sergeant Sheri Shaw

421st Fighter Squadron—Major Mark "T-Bone" Doria

AMC/PA—First Lieutenant Leslie Brown

Joint Information Bureau—Staff Sergeant Angel L. Casaigne Jr.

Nellis AFB—Captain Steven Rolenc, First Lieutenant Justin McVay, Senior Master Sergeant Charles Ramey, and Mr. Michael Estrada

General—Lieutenant Commander (Ret) Dave Parsons, Michael Jorgenson, Fred Pushies, Roberto Sanchez, and Richard VanderMuelen

Right: The unmistakable silhouette of the Batman-like B-2A stealth bomber takes to the evening skies to begin the second round of the day's Red Flag exercises. Even during nighttime missions, nothing is held back. Aircraft numbers remain the same, missions and goals are increased in intensity, and pilots are still expected to meet their respective objectives.

Chapter 1

THE HISTORY OF NELLIS AIR FORCE BASE

Red Flag. Two words that can make or break an air force fighter pilot. Red Flag is a real-world scenario designed to test pilots' skills under pressure and introduce them to their first ten combat missions. Utilizing aircraft from allied forces around the world with the Las Vegas skyline in the background, Nellis Air Force Base (AFB) plays host to Red Flag exercises many times each year.

Nellis annually provides a temporary home to an estimated 750 aircraft from 250 units located around the world. Some twelve thousand sorties and twenty-one thousand flight hours are flown with the support of eleven thousand crewmembers. Today's Red Flag exercises are among the largest and most comprehensive training programs in the world. But before going into a simulated air war, a review of Nellis Air Force Base's history can offer insights into the importance of this desert facility.

Under the direction of Colonel Martinus Stenseth, the guntruck platform was only used during a short time in January and February of 1942. The improvised moving platform put trainees on the backs of trucks to accustom them to a moving base. With the use of surplus weapons aimed at targets on railroad cars, more than nine thousand trainees graduated before the end of 1942 by becoming acclimated to aiming and firing. *USAF*

Left: Within the confines of the Red Flag building resides Suter Hall, an enormous briefing and debriefing room where pride and egos are set aside and nothing is held back. Mission objectives are addressed, performances are critiqued, and if a pilot has made a mistake during a mission, it will be identified and discussed in front of hundreds of other crewmembers. Lining the walls of Suter Hall are hundreds of plaques, awards, and commendations that testify to the intensity and importance of Red Flag operations.

In the days of the Army Air Corps, a new aerial gunnery school was needed. Army Major David M. Schlatter was assigned to find a site, and he began reviewing various locations in the Southwest United States in October 1940. On January 2, 1941, the city of Las Vegas, eager for the revenue a military base would bring, purchased a small airstrip and quickly leased what would become the Las Vegas Army Air Field to the Air Corps.

By March 1941, construction had begun on the new airfield and in May, the first base commander, Colonel Martinus Stenseth, arrived to head gunnery training. Training began in January 1942 with a collection of used weapons mounted on trucks aimed at targets on railroad cars to simulate moving targets. Due to the onset of World War II, the base's importance had increased and by the end of 1942 more than nine thousand gunners had graduated from the program. The various aircraft used for training at the time included the North American AT-6, Douglas A-33, Martin B-10, Consolidated B-24 Liberator, Boeing B-17 Flying Fortress, and Martin B-26 Marauder.

Over fifteen thousand men and women occupied the base at the height of the training period between 1943 and 1944.

As seen in 1945, the Las Vegas Army Air Field seems just as busy as the bustling base it is today—Nellis AFB. Dotting the tarmac are dozens of B-17 bombers from which aerial gunners were trained to target beefed-up RP-63 fighters with frangible bullets. Shortly after this image was taken, the base began to shift towards training gunners in the B-29 Superfortress. Two years later, the gunnery school was officially inactivated and with the creation of the USAF, the base went back to full operation as an advanced single-engine flying school in 1948. *USAF*

A derivation of the P-39 Aircobra, the RP-63 was designed to make up for the P-39's deficiencies. Seen here is a specially modified RP-63 that was used for target practice by gunnery students using special frangible rounds made of a lead/graphite composite. These rounds were designed specifically to disintegrate upon impact. The aircraft's entire standard armor was removed and over a ton of armored sheet metal was applied. The aircraft was then fitted with sensors that would detect hits signaled by illuminating a light in the propeller hub where the cannon was originally located. This earned the aircraft the unofficial nickname "Pinball." *USAF*

Although much of the training was geared toward B-17 gunners during mid-war, by 1945 most of the focus had shifted to the Boeing B-29 Superfortress. Devising a realistic target was a challenge and resulted in an innovative solution. Heavily modified Bell RP-63 fighters were developed that could be repeatedly shot at with brittle frangible bullets. Propaganda films such as *Rear Gunner,* starring Ronald Reagan and Burgess Meredith, were produced to assist with training efforts.

Towards the end of the war, the gunnery school was closed, and in January 1947 the base was officially inactivated. With the creation of the United States Air Force (USAF), the base was reactivated in March 1948 and an advanced single-engine flying school was organized. On May 2, 1949, the base held its first Air Force Gunnery Meet utilizing both jet and propeller-driven aircraft with fourteen U.S. Air Force units competing.

Las Vegas Air Force Base's new mission after reactivation was the training of pilots in advanced single-engine aircraft. Renamed in 1950 to honor a local P-47 pilot who died during the Battle of the Bulge, William Harrell Nellis, the base began training with the F-51 Mustang as well as advanced jet fighters such as the F-80 Shooting Star and the F-86 Sabre. It was the perfect location since the nearby range became an incredible resource, and Las Vegas was still relatively small, posing a minimal threat of encroachment. *USAF*

On April 30, 1950, the facility became known as Nellis Air Force Base, honoring Las Vegas resident First Lieutenant William Harrell Nellis. Nellis, a twenty-eight-year-old P-47 fighter pilot, died in action over Luxembourg during the Battle of the Bulge on December 27, 1944.

First Lieutenant William Harrell Nellis was a Las Vegas resident who died in action during the Battle of the Bulge. On December 27, 1944, Nellis was hit by ground fire while strafing a German convoy over Luxembourg. The missions undertaken by the 513th Fighter Squadron, to which Nellis belonged, saved many lives and destroyed irreplaceable German armored vehicles, personnel, and supplies. After being chosen unanimously by local civic organizations to be honored, Las Vegas Air Force Base was officially renamed Nellis Air Force Base on April 30, 1950. *USAF*

Shortly after the base's reopening, pilots began training with the North American F-51 Mustang followed by the Lockheed F-80 Shooting Stars and North American F-86 Sabres in preparation for war in Korea. A new aircraft testing and development program was also initiated at the base.

Flying high over Lake Mead, Nevada, just south of Nellis AFB, the 57th Fighter Wing shows off the various camouflage schemes worn by their aggressor F-5E Tiger II aircraft. On each aircraft resides a single AIM-9 sidewinder missile, technically known as a NATM-9M, which is equipped with special test and evaluation equipment but not a launch-capable weapon. Since live weapons were not allowed to be carried during simulated aggressor tactics and only one sensor platform was needed, it was not unusual to see these aircraft with only one store. *USAF*

B-52 Stratofortress

Air Combat Command's B-52 is a long-range heavy bomber that can perform a variety of missions. It can carry nuclear or precision-guided conventional ordnance with worldwide precision navigation capability.

For more than forty years, B-52 Stratofortresses have been the backbone of the manned strategic bomber force for the United States. The B-52 is capable of dropping or launching the widest array of weapons in the U.S. inventory. This includes gravity bombs, cluster bombs, precision-guided missiles, and joint direct-attack munitions. Updated with modern technology the B-52 continues into the twenty-first century as an important element of our nation's defenses. Current engineering analyses show the B-52's life span to extend beyond the year 2040.

The B-52A first flew in 1954, and the B model entered service in 1955. A total of 744 B-52s were built with the last, a B-52H, delivered in October 1962. Only the H model is still in the air force inventory and is assigned to Air Combat Command and the Air Force Reserves.

The first of 102 B-52Hs was delivered to Strategic Air Command in May 1961. The H model can carry up to twenty air-launched cruise missiles. In addition, it can carry the conventional cruise missile that was launched in several contingencies during the 1990s, starting with Operation Desert Storm and culminating with Operation Allied Force.

Flying with the 40th Expeditionary Bomb Squadron (EBS), a Boeing B-52H Stratofortress is loaded with joint direct-attack munitions (JDAM). The B-52 airframe is among the longest serving aircraft in the U.S. Air Force inventory and is expected to fly well into the year 2040. Capable of deploying precision weaponry, the Stratofortress maintains a heavy presence in the war on terror over Iraq and Afghanistan. *USAF*

Flying high over the Nellis Range Complex is a collection of Fighter Weapons School (FWS) aircraft as seen in 1984. Flying lead in an echelon formation is an F-111A, followed by an A-10A, a Thunderbolt II, an F-15 Eagle, an F-4 Phantom II, an F-16A Fighting Falcon, and an F-5 Tiger aircraft. It was not unusual to see every type of U.S. Air Force aircraft assembled at the base at one time for various exercises or weapons-testing applications. Today, the 57th FWS only flies the F-15, F-16, and A-10, as well as the new F-22 Raptor. *USAF*

Originally established on June 1, 1953, at Luke Air Force Base in Arizona, the USAF Thunderbirds aerobatic team relocated to Nellis AFB exactly three years later. To this day, the Thunderbirds 3600th Air Demonstration Team calls Nellis AFB their home.

In order to unite the training and research functions of the base, the Tactical Fighter Weapons Center was established in 1966. Three years later, the 57th Fighter Wing was activated in order to familiarize USAF officers with the weapons they would use in combat. Graduate-level training included air-to-air combat with both guns and missiles, air-to-ground combat tactics, as well as basic courses in fighter system maintenance and the ability to recognize faulty systems while conducting a preflight walk around.

The 64th Pursuit Squadron, originally activated in 1941, was redesignated the 64th Fighter Weapons Squadron on September 7, 1972, and activated on October 15 of that year, flying the Northrop T-38 Talon advanced jet trainer. The 65th Aggressor Squadron was activated in 1975, flying the Northrop F-5E Tiger II fighter alongside the 64th that same year. The 64th was once again redesignated 64th Aggressor Squadron on April 1, 1983.

Both squadrons were developed in response to the lessons learned from the Vietnam War. This also led to the establishment of Red Flag exercises at Nellis, which were similar in design to the U.S. Navy's Fighter Weapons School, nicknamed Top Gun. Pilots from both air force squadrons were trained on enemy tactics and engaged in mock dogfights with visiting squadrons from across the country, as well as U.S. allies.

With Nellis AFB earning a reputation for extreme combat training, a new program was added to the growing roster of exercises in 1981. Gunsmoke, a ten-day biannual gunnery meet, brought back gunnery challenges to the desert for the first time since 1962 and attracted allied forces from all over the world.

In the 1980s, Nellis AFB was in full swing. Almost every modern aircraft imaginable left rubber on its runways, including the Lockheed F-117 Nighthawk stealth fighter, which debuted at Nellis in 1988. With the Tonopah Test Range occupying part of the Nellis Air Force Range northwest of Las Vegas, this was a natural location for the Nighthawk's debut. Also occupying the range was the Indian Springs Air Force Auxiliary Airstrip. This little-known part of the test range was home to the 11th, 15th, 17th, and 30th Reconnaissance Squadrons, which operate the General Atomics Predator RQ-1, MQ-1 and MQ-9 unmanned aerial vehicles (UAVs). In February 2001, a Predator UAV successfully test fired its first Hellfire missile over the Nellis test range. The airstrip was renamed Creech Air Force Base on June 20, 2005.

The various components that make up Nellis Air Force Base include the Air Warfare Center (AWFC), responsible for graduate-level air-combat training for mission-ready aircrews

C-130 Hercules

The C-130 Hercules primarily performs the tactical portion of the airlift mission. The aircraft is capable of operating from rough dirt strips and is the prime transport for air dropping troops and equipment into hostile areas. The C-130 operates throughout the U.S. Air Force, and fulfills a wide range of operational missions in both peace and war situations. Basic and specialized versions of the aircraft airframe perform a diverse number of roles, including airlift support, Antarctic ice resupply, aeromedical missions, weather reconnaissance, aerial-spray missions, firefighting duties for the U.S. Forest Service, and natural-disaster-relief missions.

The Lockheed C-130 Hercules is the most common transport aircraft in the West and has been in production longer than any other aircraft in history. The prototype flew in August 1954, and since then over sixty nations have ordered the Hercules. The initial production model was the C-130A, with a total of 219 units ordered. Deliveries began in December 1956.

Introduced in August of 1962, the 389 C-130Es added two 1,290-gallon external fuel tanks and an increased maximum takeoff weight capability.

The latest C-130 to be produced, the C-130J, entered the inventory in February 1999. With the noticeable difference of a six-bladed composite propeller coupled to Rolls-Royce AE2100D3 turboprop engines, the C-130J brings substantial performance improvements over all previous models, and has allowed the introduction of the C-130J-30, a stretch version with a fifteen-foot fuselage extension. To date, the air force has taken delivery of thirty-seven C-130J aircraft.

The latest model in the C-130 line is the Lockheed C-130J Hercules. Having just celebrated fifty years of service, the C-130 airframe is still rolling off assembly lines. The versatile four-engine turboprop has seen use by countries all over the world and has been indispensable in transporting materials needed for Operation Enduring Freedom. The Hercules has been subjected to almost every condition from experimental carrier landings aboard the USS *Forrestal* in the 1960s to jet-assisted takeoffs (JATO) from the dirt strips of Vietnam. For Red Flag exercises, the Hercules can be tasked with multiple roles including electronic surveillance, airborne tanker, and even remote vehicle delivery for advancing ground troops.

throughout the world. Additionally, the headquarters of the AWFC conducts follow-on operational testing and tactics development and evaluation using the latest weapons systems that equip U.S. air forces for combat. The operational elements of the center are the 57th Wing at Nellis, the 99th Air Base Wing at Nellis and the 53rd Wing from Eglin AFB, Florida.

The 57th Wing provides advance training for the combat air force. The wing is comprised of the 57th Operations Group, 57th Logistics Group, U.S. Air Force Weapons School, U.S. Air Ground Operations School (AGOS), and the U.S. Air Force Air Demonstration Squadron—the Thunderbirds. The groups provide the logistics and operations infrastructure to conduct advanced combat training at unit and individual levels. The wing also has

A fully loaded F-15E Strike Eagle from the 57th Wing Air Force Weapons School is towed across the tarmac. The mission of the USAF Weapons School (USAFWS) is to teach graduate-level instructor courses, which provide the world's most advanced training in weapons and tactics employment to officers of the combat air forces. Besides the Strike Eagle, the USAFWS operates the OA-10A Thunderbolt II, the F-15C Eagle, and the F-16C Fighting Falcon at Nellis AFB. Other aircraft such as the B-52, B-2, B-1, HH-60G, AC-130, F-117, and MH-53, Command and Control Operations, Intelligence, Space, and Support are operated at other USAFWS attachments throughout the country.

Above: Assigned to the 66th Rescue Squadron (RQS), an HH-60G Pave Hawk helicopter flies high above the Nellis Test and Training Range. With the range consisting of such harsh and diverse conditions, it was only natural that a component of Red Flag operations consists of an elaborate search-and-rescue (SAR) exercise. Once the downed pilot has been found by Blue Air and the local area deemed safe for rescue, the 66th RQS performs a standard SAR insertion to recover the downed aviator and return him to friendly territory. *USAF*

KC-135 Stratotanker

The KC-135 Stratotanker's principal mission is air refueling. This unique asset greatly enhances the USAF's capability to accomplish its primary missions of global reach and global power. It also provides aerial refueling support to air force, navy, and marine aircraft, as well as aircraft of allied nations.

Boeing's Model 367-80 was the basic design for the commercial 707 passenger plane as well as the KC-135A Stratotanker. In 1954, the air force purchased the first twenty-nine of its future 732-plane fleet. The first aircraft flew in August 1956 and the initial production Stratotanker was delivered to Castle Air Force Base, California, in June 1957. The last KC-135 was delivered to the air force in 1965.

Through the years, the KC-135 has been altered to do other jobs ranging from flying command-post missions to reconnaissance. RC-135s are used for special reconnaissance and Air Force Material Command's NKC-135As are flown in test programs. The Air Combat Command operates the OC-135 as an observation platform in compliance with the Open Skies Treaty.

Over the next few years, the aircraft will undergo upgrades to expand its capabilities and improve its reliability. Among these are improved communications and navigation and surveillance equipment to meet future civil air traffic control needs.

A Pittsburgh KC-135 from the 171st Aerial Refueling Wing shows off its nose art with a tanker from the 141st ARW based at Fairchild, Washington, quietly sitting on the ramp. For each Red Flag sortie, a tanker is assigned to either Red Air or Blue Air to replenish fighters out on the Nellis Range. Typically, two tankers are provided for Blue Air and one tanker for Red Air due to the respective size of the forces.

several operational units consisting primarily of UAVs located at nearby Creech AFB and Sikorsky HH-60 Pave Hawk helicopters from the 66th Rescue Squadron (RQS).

The primary mission of the 66th Rescue Squadron is worldwide combat rescue in support of combat air forces. The 66th RQS is one of only four active-duty air force HH-60 combat rescue units and is geared for worldwide deployment. The 66th RQS performs other vital functions, including rescue support for air operations over the Nellis Range Complex and backup rescue for civilian agencies in the local area and the greater southwestern United States. Depending on the mission, a typical rescue crew includes a pilot, copilot, flight engineer (FE), and two pararescue jumpers (PJs). PJs are qualified combat paramedics, scuba divers, parachutists, mountain climbers, and survivalists.

The 422nd Test and Evaluation Squadron (TES) is composed of aircrew and support personnel who are responsible for five different flights of fighter and helicopter aircraft: Fairchild Republic A-10, McDonnell Douglas F-15C, F-15E, General Dynamics/Lockheed F-16C, and HH-60G. The 422nd TES conducts operational tests for the U.S. Air Force's Air Combat Command (ACC) on new hardware and upgrades to each of the five aircraft in a simulated combat environment. The 422nd TES also develops and publishes new tactics for these aircraft. The results of these tests directly benefit aircrews in ACC, Pacific Air Forces (PACAF) and United States Air Force in Europe (USAFE) by providing them with operationally proven hardware and software systems. Current tests include employment of the AGM-65H Maverick missile by the A-10, developing night-vision goggle employment tactics for the F-16C, helmet-mounted sight capability for the F-15C, new avionics software updates for the F-15E, and improved combat search-and-rescue tactics for the HH-60G. In 2003, Nellis was

Left: Along with the 53rd Fighter Wing stationed at Eglin AFB, Florida, which reports to the Air Warfare Center at Nellis, the 33rd Wing is also based at Eglin and operates two squadrons, the 58th and the 60th. Seen here is the flagship F-15C from the 33rd Fighter Wing launching off runway 3L. The 33rd Fighter Wing became known as the "Nomads" due to their extensive worldly assignments dating back to World War II and their missions over the European and Pacific theaters.

Home of the fighter pilot, Nellis Air Force Base is one of the busiest bases in the world. Calling Nellis its home is the 57th Fighter Wing, which is responsible for the 547th Intelligence Squadron and the 414th Combat Training Squadron. Also located at Nellis are the Joint Air Ground Operations Group (JAGOG), the 57th Maintenance Group (MXG), the USAF Weapons School (USAFWS), the USAF Thunderbirds, the 99th Air Base Wing (ABW), the 98th Range Wing, and the 422nd Test and Evaluation Squadron (TES).

selected for the F-22 Force Development Evaluation Program where the 422nd became home to the first of seven production Lockheed Martin F-22 Raptor fighters.

The 99th Air Base Wing (ABW) provides command guidance for all support agencies located at Nellis or associated with the center. The wing is separated into three groups: 99th Support Group, 99th Logistics Group, and 99th Medical Group. Each group provides a wide array of services including transportation, supply, morale, welfare, recreation and services, contracting, civil engineering, security police, mission support, and communications. It also provides the base's aerospace medicine, and hospital, dental, and nursing services. The wing is responsible for the operation of the new 114-bed Air Force/Veterans Administration (VA) hospital located outside Nellis' North Gate on Las Vegas Boulevard, North.

The 53rd Wing located at Eglin AFB, Florida, reports to the Air Warfare Center at Nellis, a direct reporting unit (DRU) to Headquarters ACC. It serves as the focal point for the combat air forces in electronic combat, armament and avionics, chemical defense, reconnaissance, command-and-control, and aircrew-training devices.

Tenant units at Nellis include the Defense Commissary Agency (DCA); the 820th Rapid Engineer Deployable Heavy Operational Repair Squadron Engineer (RED HORSE) Squadrons; the 372nd Training Squadron, Detachment 13; the Air Force Audit Agency (AFAA), Detachment 250; and the 896th Munitions Squadron. Other Nellis tenant units include the USAF Area Defense Counsel (ADC); the Air Force Office of Special Investigations (AFOSI), Detachment 206; the Defense Reutilization and Marketing Office (DRMO); and a federal prison camp (FPC).

Nellis' Area II is an integral part of the base and is located at the northeast edge of the main base. In September 1969, this area became part of the Nellis complex. Previously, it served as a

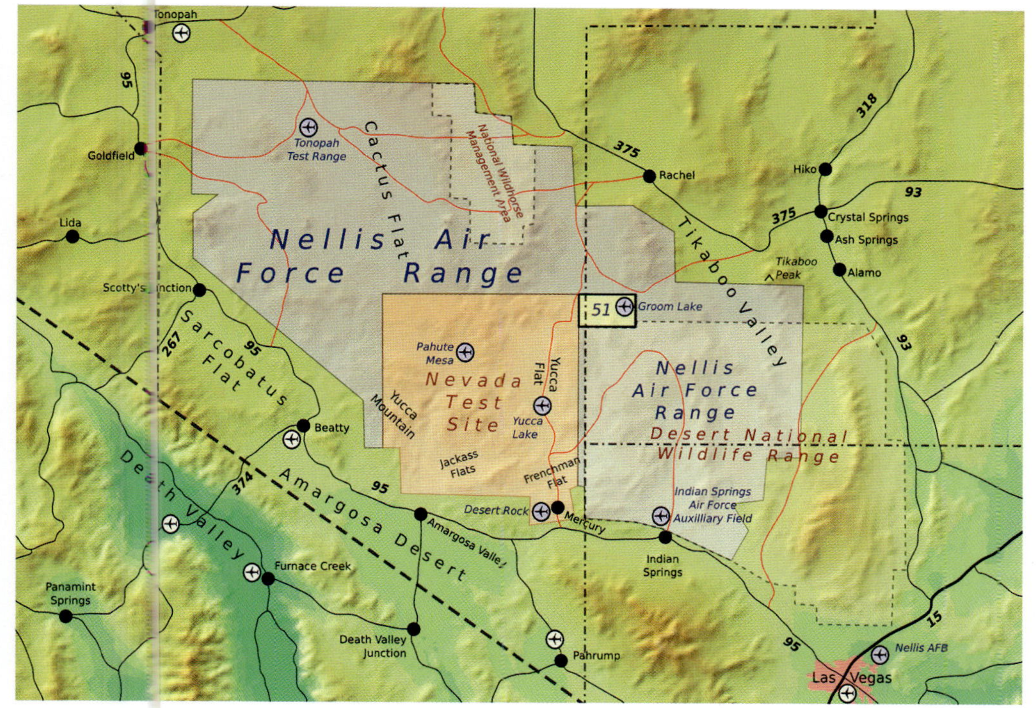

The Nellis Range Complex (NRC) is the largest area of land and controlled military airspace in the lower forty-eight states. Although most of the land is owned and operated by the Department of the Interior (DOI) and the Bureau of Land Management (BLM), the range is also maintained by the 98th Range Wing stationed at Nellis Air Force Base. Consisting of nearly twelve thousand square miles, the NRC is second only to the newly created Red Flag–Alaska, which operates within an enormous sixty-six-thousand-square-mile radius. Within the NRC is the off-limits Groom Lake Area 51 facility that military pilots are forbidden to over fly despite their respective missions. Because of the Area 51 restrictions and its central location, many pilots compare the range to flying within a doughnut, which adds another degree of difficulty.

weapons storage area for the United States Navy. Three units are based in this area: the 57th Equipment Maintenance Squadron (EMS), which provides safe and reliable munitions handling in support of the combat mission; the 896th Munitions Squadron, part of Air Force Materiel Command; and the 820th Red Horse Squadron.

The Department of Energy (DOE) installation is located in Nye County, approximately sixty-five miles northwest of Las Vegas with support and administrative headquarters at Mercury, Nevada. It is operated by the DOE, Nevada Operations Office in Las Vegas, which is charged with the management of all the nation's nuclear detonation programs, including this area known as the Nevada test site.

The Nevada test site covers approximately 1,350 square miles. It includes the Yucca and Frenchman Flats, Paiute and Rainier Mesas, and the former Camp Desert Rock area, which was used by the Sixth Army in the 1950s to house troops participating in atmospheric tests at the test site. Yucca Flat, a valley roughly 10 miles wide by 20 miles long, and Paiute Mesa, a rugged 7,500-foot-high area of 166 square miles at the northwest corner of the site, are the main underground test areas.

Frenchman Flat is the first dry lake basin north of the hills beyond Mercury. It was used for most of the nuclear blast tests between January 27, 1951, and March 25, 1968, but has since been used primarily for DOE weapons development tests and Department of Defense military effects tests.

Today, Nellis Air Force Base continues to lead as one of America's most advanced military installations, covering nearly eleven thousand acres with approximately ten thousand employees. An integral part of ACC, Nellis AFB is known as the "Home of the Fighter Pilot."

Nellis Range Complex

What makes Nellis AFB so valuable and the Red Flag exercise so successful is the Nellis Air Force Range (NAFR) or Nellis Range Complex (NRC). The range contains the largest area of land and controlled military airspace in the continental United States with weather that is reasonably predictable and suitable for year-round flying.

This enormous amount of land encompassed nearly 3,560,000 acres when established by President Franklin D. Roosevelt in 1940. Originally referred to as the Las Vegas Bombing and Gunnery Range, Executive Order 9019 returned approximately 937,730 acres to the authority of the Department of the Interior (DOI) in 1942. Five years later, the Tonopah Bombing and Gunnery Range turned over an additional 154,584 acres to the DOI. After a few more instances of trading back and forth with the DOI and the Bureau of Land

One of the many targets located throughout the Nellis Range Complex is the remainder of this F-4 Phantom II. Although the Bureau of Land Management and the Department of the Interior own much of the land, the 99th Mission Support Group maintains the Nellis Range Complex. Targets such as these are not only used for Red Flag operations but also for gunnery exercises of the A/O-10A Thunderbolt II. Along with dilapidated airframes and target vehicles, runways and airbases can quickly be created to simulate what pilots may find when over hostile territories. *USAF*

Management, the Nellis Air Force Range, more formally known as the Nevada Test and Training Range (NTTR), currently consists of approximately 2.9 million acres of land. The airspace over an additional five million acres is shared with commercial aircraft encompassing the Nellis Range Complex.

Operating and maintaining the twelve thousand square miles of airspace and land that make up the complex is the job of the 98th Range Wing and the 99th Mission Support Group.

The 98th Range Wing (RANW) operates, maintains, and develops the Nevada Test and Training Range including two local airfields, Creech AFB and the Tonopah Test Range, as well as the instrumentation for Air Warrior at the National Training Center (NTC) and Leach Lake Range. The 98th also works closely with the Department of Defense in support of advanced composite force training tactics development and electronic combat testing. Together with the DOD and the Department of Energy, the 98th further pursues testing requirements and research and development procedures.

Part of the 99th Air Base Wing, the Mission Support Group consists of six different squadrons: 99th Communications Squadron, 99th Civil Engineer Squadron, 99th Mission Support Squadron, 99th Contracting Squadron, 99th Services Squadron, and 99th Logistics Readiness Squadron. Detachment One of the 99th Range Group provides support to the southern portion of the Nellis Range Complex as well as Creech AFB. Detachment Two directs all ACC activities at the Tonopah Test Range Airfield and the Northern Ranges. Both detachments provide support for

Sorting out the sectors of the Nellis Range Complex is vital for safety and efficiency over the range. A-10 Warthog pilots Captain Matt McGarry and Captain Matt Vilalla brief flight-plan coordinates with Captain Brett Gallagher prior to flying a 2002 Red Flag mission at Nellis AFB. A-10 Warthogs provide essential ground support removing threats that could jeopardize air defense forces or advancing friendly troops. *USAF*

Throughout the Nellis Range Complex (NRC), missile sites, antiaircraft guns, and simulated runways dot the landscape such as this simulated SCUD missile protruding from a mound. Obstacles on the range are designed by members of the 414th Combat Training Squadron (CTS), formerly known as the 4440th Tactical Fighter Training Group, to simulate those encountered by fighter aircraft in real-world combat operations. Functional simulated technology equipment located on the range such as surface-to-air missiles and acquisition radar sites are operated by the 39th Intelligence Squadron. *USAF*

Allied Aircraft: Hercules C models

C-130Ks in use by the RAF are known as Hercules C1s and C3s, and initial deliveries (of a total of sixty-six ordered) were made during the mid-1960s. Many are destined to remain in service for some years to come, although the RAF has replaced some of its older Hercules aircraft with second-generation C-130Js on a one-for-one basis. Twenty-five of these new aircraft, known as the Hercules C4 and C5, were ordered in December 1994, with deliveries commencing in 1998.

The fleet is based at RAF Lyneham in Wiltshire and operated by crews from No. 47 and 70 Squadrons—No. 24 and 30 having changed to the C4 and C5 versions. One aircraft is also based at Mount Pleasant Airfield in the Falkland Islands with 1312 Flight. The Hercules is used primarily to carry personnel and support equipment, and to airlift patients to home base or to other allied countries.

No single country precedes allied participation more than the United Kingdom. Rolling down runway 27L is a C-130K flown by the No. 47 Squadron out of Lyneham. The No. 47 Squadron has been flying the C1 and C2 Hercules variants since 1968 and has been based out of Lyneham since 1971. UK aircraft have been designed to incorporate the drogue aerial refueling method and, as such, British C-130s as well as most UK aircraft, have long probes extending forward the fuselage.

recovery of emergency or diverted military aircraft during the various exercises.

With direct relation to Red Flag exercises, the 99th is responsible for scoring sites at Belle Fourche, South Dakota; La Junta, Colorado; Dugway, Utah; and Harrison, Arkansas, as well as an instrumentation-support facility located at Ellsworth AFB, South Dakota. The Nellis Range Complex maintains instrument support facilities and two emergency-divert airfields and they work closely with the Department of Energy, Department of the Interior, Bureau of Land Management, and, of course, the Department of Defense to meet a broad spectrum of range user requirements.

Since the NRC had been designated a major range and test facility by the Department of Defense, the 99th Range Group now acts as the Air Combat Command lead range advocate to provide centralized expertise for the development of ACC test and training ranges. The Range Group operates with the assistance of approximately 600 contractors and nearly 300 military and civil service personnel.

The Nellis Range Complex supports numerous Red Flag and Green Flag exercises along with multiple USAF Weapons School exercises each year. The NRC also hosts the Gunsmoke competition every two years. Operational testing and evaluation missions by the 422nd Test and Evaluation Squadron on the NRC are supported by upgraded television ordnance scoring systems (TOSS) and state-of-the-art Kineto tracking mount documentation and time-space position information (TSPI) data. Additional capabilities include support for operational flight programs (OFP), qualification operational test and evaluation (QOT&E), tactics development and evaluation (TD&E), and follow-on test and evaluation (FOT&E).

The training range maintains some of the most realistic integrated threat simulator technology in the world. In addition to the assortment of surface-to-air missiles (SAMs), antiaircraft artillery (AAA), and acquisition radars operated by personnel from 39th Intelligence Squadron, they also maintain and operate a variety of radar and communications jamming equipment. Coupled with the Nellis Red Flag measurement and debriefing system (RFMDS), these assets provide superior year-round training to U.S. and allied aircrews in both competition and training exercises.

Should real-world circumstances require additional realistic configurations for training, targets can be built or modified quickly. In one example, range contractors transformed

A Block 15 F-16A, serial 80-3598 from Denmark takes to the skies during a 2004 Red Flag exercise. Eskadrille 726 based at Aalborg, Denmark, transitioned to the F-16 from the F-104G in late 1983. Denmark was one of the original four countries to bring the F-16 to Europe. Both the United States and allied nations benefit from the cross training obtained from Red Flag exercises. This particular exercise gave the Danes the opportunity of testing their newly equipped operational satellite-guided technology, which all of their F-16s will eventually include.

a runway configuration from a typical former Warsaw Pact country's layout to one based on what allied aircrews would see in Iraq using data gathered from intelligence reports and photo reconnaissance missions.

The Nellis Range Complex is one of the most versatile and hazardous ranges in the United States, the perfect environment for honing search-and-rescue skills. Although providing rescue support for air operations over the range is the 66th Rescue Squadron's secondary mission, its importance is none the less vital to range exercises.

The NRC is located between Las Vegas and Tonopah in southwestern Nevada and consists of five adjacent geographical areas. Those areas include the restricted areas R-4806, primarily used for testing and munitions training; R-4807, used for electronic combat and munitions training; R-4808, used by the Nevada Test Site; R-4809, used primarily as an electronic combat range; and the Desert Military Operating Area, used for air-to-air combat training. The land throughout the complex is mostly barren, consisting of dry washes and lakebeds along with rugged, mountainous terrain and typical desert vegetation.

Much of the complex is comprised of land withdrawn from the Bureau of Land Management, and it's off-limits to the general public. However, portions of the range are set aside for

A couple of A-10 Thunderbolt IIs fly high over the barren Nellis Range Complex (NRC) after taking on fuel from a KC-135R tanker. Belonging to the 52nd Fighter Wing, 81st Fighter Squadron from Spangdahlem Air Base in Germany, these A-10s have plenty of room to play. With most of the NRC being off-limits to the general public, A-10s, along with other low-flying aircraft like the HH-60G's belonging to the 66th Rescue Squadron, can practice map-of-the-earth low-terrain flying with no risk to public safety.

livestock grazing and the Nevada horse range. Located directly in the center of NRC is the controversial and highly secretive Groom Lake/Area 51 complex which is even off-limits to those using NRC for combat training scenarios.

Creech Air Force Base

North of Nellis Air Force Base, just beyond the outskirts of Las Vegas, Nevada, lies Creech Air Force Base. Part of the Nellis Test and Training Range (NTTR) or Nellis Range Complex (NRC), Creech AFB is home to the unmanned aerial vehicle battle lab, 11th Reconnaissance Squadron, and Silver Flag Alpha, a desert warfare training center that trains over three thousand security force personnel each year.

Originally constructed by the army to support World War II efforts, it wasn't until 1964 that the airfield was named the Indian Springs Air Force Auxiliary Field after the small nearby community bearing the same name.

On June 20, 2005, Indian Springs Air Force Auxiliary Field was officially changed to Creech Air Force Base in honor of General W. L. "Bill" Creech. By advocating a philosophy of decentralized authority and responsibility, General Creech revolutionized the air force through his leadership as the commander of Tactical Air Command. A fighter pilot and war hero of Korea and Vietnam, he improved many tactics that led to the success of air missions in the Persian Gulf, Kosovo, Iraq, and Afghanistan conflicts. It has been said that no single officer has had a greater impact on the U.S. Air Force in recent years.

In developing the base for use with medium-altitude endurance unmanned aircraft, the talents of three districts came together on a single project for the first time in South Pacific Division history. The Army Corps of Engineers' Albuquerque, Los Angeles, and Sacramento districts aimed to develop a "One Door to the Corps" concept. Once Sacramento had completed approximately 35 percent of the project, it was then "brokered" to the Albuquerque district to complete the design and advertise the project. The Los Angeles district supplied open bids, project management, and awarded and administered the contract, while the Sacramento district furnished technical review and project support.

Both the 11th and 15th Reconnaissance Squadrons operate the Predator RQ-1A/B systems in response to the DoD requirement to provide improved intelligence, reconnaissance, and surveillance information to the battlefield. After the USAF was selected as the operating service for the RQ-1A system, the 11th RS acquired its first two Predator UAVs in November 1996.

Unmanned Aerial Vehicle Battlelab

The mission of the unmanned aerial-vehicle battlelab (UAVB) is to work with combatants and identify problems that can be solved using innovative remotely piloted UAV aircraft. The UAVB team works with industry, academia, and federal government and military labs to find solutions to combat challenges, demonstrate those solutions, make an objective decision about their military utility, and work to transition the initiative to the battlefield. To accomplish that mission, UAVB pursues high-return initiatives with minimal investment and maximum impact for UAV combat organizations, training, and future requirements and acquisitions.

The UAV battlelab intelligence, surveillance, and reconnaissance (ISR) division is led by Major Matthew Belmonte. The division analyzes existing and emerging technologies for integration into joint UAV systems with effects-based emphasis

Members of the 11th Reconnaissance Squadron at Creech AFB, Nevada, perform preflight checks on the RQ-1 Predator prior to a mission. The 11th and 15th Reconnaissance Squadrons support, test, and train pilots and crewmembers in the operation and continued development of the RQ-1 systems. Like the RQ-4 Global Hawk UAV, the Predator has participated very little in past Red Flag exercises, though heightened expectations of combat realism will allow the Predator increased involvement in the near future. *USAF*

RC-135V/W Rivet Joint

The RC-135V/W Rivet Joint reconnaissance aircraft supports theater- and national-level consumers with near-real-time on-scene intelligence collection, analysis, and dissemination capabilities.

The current RC-135 fleet is the latest iteration of modifications to this pool of C–135 aircraft going back to 1962. Initially employed by Strategic Air Command to satisfy nationally tasked intelligence collection requirements, the RC-135 fleet has also participated in every sizable armed conflict involving U.S. assets during its tenure. RC-135s were present supporting operations in Vietnam, the Mediterranean for Operation El Dorado Canyon, Grenada for Operation Urgent Fury, Panama for Operation Just Cause, and Southwest Asia for operations Desert Shield, Desert Storm, Enduring Freedom, and Iraqi Freedom. RC-135s have maintained a constant presence in Southwest Asia since the early 1990s.

All RC-135s are assigned to Air Combat Command. The RC-135 is permanently based at Offutt Air Force Base, Nebraska, and is operated by the 55th Wing, using various forward-deployment locations worldwide.

Stationed at Offutt Air Force Base in Nebraska, this RC-135V/W Rivet Joint electronic surveillance aircraft is from the 55th Wing, 38th Reconnaissance Squadron. Sometimes referred to as "hogs" due to the extended nose and prominent cheeks, the RC-135 is capable of providing indications about the location and intentions of enemy forces as well as warning friendly forces of threatening activity. There are currently fifteen RC-135s in service and all are based at Offutt AFB. The flight crew is trained by the 95th Reconnaissance Squadron and the intelligence personnel by the 488th Intelligence Squadron, both at the Royal Air Force Base in Mildenhall, England.

on ISR operations, to include radar, video, laser-designation, and other ISR systems. The division manages evolutionary and revolutionary initiatives demonstrating enhanced UAV ISR capabilities. Each initiative is focused on moving successfully demonstrated technology from the battlelab to the combat field. Finally, the ISR division serves as an advisory group to ACC and air staff on UAV ISR issues.

The UAV Battlelab Combat Applications Division is led by former *Thunderbird* team member and fighter pilot Major Doug Larson. This UAV Battlelab Combat Applications Division analyzes existing and emerging technologies for integration into joint UAV systems with effects-based emphasis on kill-chain operations, to include weaponization, sensor delivery, and electronic warfare systems. With this in mind, the division manages innovative initiatives to demonstrate enhanced UAV attack and targeting capabilities. Incorporated into the initiative is a detailed roadmap to successfully move from the battlelab to acquisition or operational status. These initiatives are then transitioned to UAVs in a combat environment in support of the warfighter. The Combat Applications Division also serves as an advisory group to ACC and air staff on UAV weapon-related issues.

Battlelab Projects

The tactical UAV coordinator workstation—and its next-generation version, the UAV battle-management system—is used by the USAF UAVB and army space and missile defense

Utilizing the RQ-1 Predator, the unmanned aerial vehicle (UAV) battlelab at Creech AFB works on integrating existing technologies into the UAV platform. Besides surveillance applications, the UAVB continues to apply enhanced UAV attack and targeting capabilities to airframes such as the Predator. The goal of the UAVB is to devise solutions with minimal cost and maximum impact in the development of combat-capable UAVs. *USAF*

battlelab to demonstrate state-of-the-art tools for managing a variety of UAV missions, including surveillance and reconnaissance, combat search and rescue, hunter-killer or killer-scout missions, and forward air control. The single workstation combines UAV mission data, visualization tools, intelligence data, communication tools, and intelligent agent applications, allowing the tactical UAV coordinator to quickly retrieve data to support tactical commander decisions, maximizing the efficiency of UAVs. The TCW aids operators with mission coordination, monitoring, and managing, supporting both the Air Operations Center (AOC) and the individual ground control stations (GCS).

The air traffic detection sensor system (ATDSS) is a passive moving-target detection technology. The Air Force Research Laboratory Sensors Directorate (along with the Predator and Global Hawk program offices) sponsored Defense Research Associates (DRA) to adapt missile-detection technology to use a see-and-avoid application. ATDSS uses DRA's proprietary software and low-cost optical sensors and processors to detect collision-course aircraft.

The Little Weasel initiative will use electronic intelligence (ELINT) capability on a UAV to support lethal suppression of enemy air defenses (SEAD). Little Weasel integrates battle-proven technologies—including the HARM (antiradar missile) targeting system pod and the improved data modem (IDM)—to provide the remote capability to detect and identify enemy threat emitters. Combined with a long-endurance platform, this will provide persistent battlefield ELINT that can pass SEAD messages directly to fighting aircraft or other SEAD platforms, which will then attack the target.

Although the potential for UAVs to provide intelligence, surveillance, and reconnaissance (ISR) coverage for convoy support has long been recognized by the military, assets like Predator or Global Hawk have limited availability. The urgent need for persistent ISR data for convoy support has made the development of small, low-cost UAVs a priority. The UAVB—collaborating with the force protection battlelab and the command-and-control battlelab—has met the performance requirements of the 820th Security Forces Group and the 732nd Expeditionary Mission Support Group with the Boeing ScanEagle A-15. The unpiloted aircraft increases the convoy commander's situational awareness, detects improvised explosive devices (IEDs), and complements air base defense and mobile patrols within the area.

Chapter 2

THE ORIGINS OF RED FLAG

For each step of the way toward a pilot's proficiency in mastering the skills of precision flying, there is always some form of training. From physical fitness to flight simulation, every aspect of flying is slowly introduced to the aspiring combat pilot. But how do you simulate the stress, confusion, tactics, and skills involved in actual combat?

Red Flag, Green Flag, Blue Flag, Hawgsmoke, Air Warrior II, and more are just some of the exercises held over the years at Nellis Air Force Base. The most prominent— Red Flag—is designed to provide an accurate representation of the combat experience. Utilizing the latest in available technology and adversary warfare techniques, Red Flag participants come as close as humanly possible to experiencing a combat environment.

Shortly after World War II and into the Korean War, the United States Air Force began to rely heavily on strategic bombing efforts. An enormous accountability was placed on the

Above: Utilizing a 1960s Boeing 707 airframe, describing the E-3 Sentry (AWACS) aircraft as being a slightly modified variant would be a gross understatement. At an individual cost of about $270 million, the Sentry consists of a flight crew of four plus a mission crew of thirteen to nineteen specialists depending on the mission. As an airborne surveillance, command, control, and communications aircraft, the E-3 can scan from the Earth's surface up into the stratosphere, over land or water for both friendly and enemy forces.

Left: Staring down the length of runway 3L, four F-15C Eagles can be seen in this unusual view prior to a nighttime mission launch. Within seconds, each F-15C will engage afterburners and take off one by one. Once airborne they will join up in an echelon or fingertip formation on their way out to the range. For nighttime missions they will return one by one on a direct approach as opposed to daytime missions where they will return in echelon formation and conduct an overhead break prior to landing.

Proving to be a formidable adversary during the Korean and Vietnam Wars, the MiG-15 and MiG-17 aircraft were extremely nimble and took full advantage of the weaknesses of U.S. forces aircraft such as the F-4 Phantom II. With the U.S. concentrating on beyond-line-of-sight tactics, as well as a heavy bombing role, dogfighting skills began to suffer. MiGs would simply intercept the fighters and force them to dispose of their loads prematurely, creating a dogfight for which American pilots were not appropriately trained. As the statistical kill ratio dropped to an unfavorable figure, the creation of Red Flag to properly train pilots in combat tactics within a safe environment was born.

Strategic Air Command (SAC) as well as precision delivery of munitions by fighter aircraft. Air-to-air scenarios and close-range combat, along with efficient dogfighting techniques began to falter. With the introduction of the Soviet-built MiG-15 fighter during the Korean War, the USAF soon found its aerial dominance fading fast. With kill ratios heavily favoring the United States during World War II, results of the Korean War dropped to an acceptable, but unofficial, ten to one (ten enemy losses to one U.S. loss). This failed to cause any concern or initiate an interest in reviving more extensive air combat tactics training.

With the Korean War in the history books, the Vietnam War began to escalate and U.S. involvement increased. Unlike the Korean War, where the enemy routinely initiated combat, the North Vietnamese air forces commonly left American combat air patrol fighter aircraft alone and only challenged fighter-bombers, forcing them to prematurely dispose of their munitions in order to evade enemy combatants. This new agenda by the North Vietnamese helped protect potential targets such as airfields and weapons depots and decimated U.S. efforts to advance the war in their favor. With many U.S. aircraft designed for bombing strategic targets, it was common to encounter sophisticated defense systems, surface-to-air missiles, and antiaircraft artillery as well as enemy aircraft being assisted by enhanced ground control.

This frustrated USAF and U.S. Navy forces to the point where a primary tactic was looking for a fight. Since bombing efforts were proving unsuccessful and certain rules of engagement prohibited the attack of MiG airbases, the only way to rid the skies of enemy aircraft was by air-to-air combat. Much of that combat was against the MiG-17 fighters, the successor to the Korean-era MiG-15. Though much was learned from the defected MiG-15, the North Vietnamese (and Soviet) pilots knew how to handle the MiG-17s by utilizing optimum aerial tactics and taking full advantage of the fighters' altitude and speed characteristics.

Besides failed missions, U.S. forces were dealt a major blow when calculating kill ratios around the beginning of 1972. Recording at times less than one to one, something needed to be done.

The U.S. Navy first introduced Top Gun for naval aviators as an instructional school based on flying against dissimilar aircraft. The USAF continued to have impromptu training sessions involving similar aircraft—essentially training pilots for combat scenarios that they would never face.

Although the McDonnell Douglas F-4 Phantom II was a potent adversary compared to the various Communist aircraft, the flight characteristics were vastly different. Not only was the Phantom II big, but it was a multi-role aircraft with relatively poor visibility. The roll rate, climb rate, turning capabilities, and other aspects of the more nimble MiGs simply couldn't be duplicated by the Phantom. Even worse, pilots flying the adversary role were not specifically trained in the combat tactics of Communist air forces and often belonged to the same units as those they were flying against.

Noticing these deficiencies, Colonel John Boyd developed various aerial maneuverability concepts regarding the new F-4E and its improved flight characteristics. However, the techniques and proficiency took some time to adopt. The Achilles' heel of Boyd's concepts was the aircraft itself along with the limiting rules of engagement. Armaments of the F-4E were limited to AIM-7 Sparrow and AIM-9 Sidewinder missiles, which were intended to engage larger bombers and were ineffective against the more nimble aircraft performing extreme maneuvers. Because of the conditions of the Vietnam War, visual identification of enemy aircraft was required, resulting in close air combat where missiles were essentially unusable. The adversary's aircraft relied heavily on guns and were quickly

During a 1987 Red Flag exercise, Captain James Stanton and First Lieutenant Miguel Nieves from the 355th Tactical Fighter Squadron run through their respective checklists in an F-4E Phantom II. Although the F-4E included improved flight characteristics, the aircraft still had poor visibility and limited maneuverability compared to the more nimble MiG fighters. Colonel John Boyd took on the tasks of devising new maneuverability concepts taking full advantage of the improved flight characteristics. *USAF*

Allied Aircraft: VC10

In RAF service since 1966, the VC10 comes in two versions. The VC10 C1K has three roles:

Air Transport—The movement of personnel and support equipment.

Air-to-Air Refueling—The facility to extend the range, endurance, or payload of air operations.

Aeromedical Evacuation—Transport of patients (usually via stretcher).

The VC10 C1K, although now old and suffering from both limiting airframe restrictions as well as air traffic restrictions, such as noise and the new avionics requirements for modern civil airspace, still remains a very capable aircraft. The ability to rapidly deploy fast-jets and carry ground personnel or support equipment worldwide keeps it at the heart of the flexibility of the RAF. Although hardly ever in the spotlight, the aircraft is a true workhorse of the RAF and has been involved in the majority of conflicts in the last thirty-five years or so.

The newer VC10 K3s and K4s form the bulk of the RAF's air-to-air refueling fleet. A replacement for both versions of the VC10 and the Tristar is being sought under the future strategic tanker aircraft (FSTA) program. It was announced in January 2004 that the AirTanker consortium, which was offering Airbus A330 aircraft, had been chosen as the preferred partner for the FSTA contract.

One-Zero-One Squadron is the only active UK aerial tanker unit operating the VC10 C1K. Seen here flying over Oxfordshire, England, aircraft XR810 shows off its wing-tip- mounted refueling pylons from which a tethered drogue is deployed. The VC10 C1K aircraft functions as a dual-role airframe not only taking on the tasks of an aerial refueler but also that of cargo and personnel transport. The aircraft can carry up to 154,000 pounds of fuel utilizing its original eight fuel tanks and can either be used to feed the aircraft itself or be dispensed to smaller fast-jet receivers. *Michael Jorgenson—www.actionairimages.com*

Taxiing down the Nellis flightline, an F-4E Phantom II aircraft assigned to the 474th Tactical Training Wing looks for a spot to park at the conclusion of a mission. After relocating to Nellis AFB, Nevada, on June 21, 1975, the squadron transitioned to the F-4E and became tasked with conducting fighter training. The Phantom brought with it a multitude of capabilities, including improved bombing and aerial adversary qualities. It was up to the 474th to maximize a pilot's proficiency in mastering both roles. *USAF*

taking the upper hand. More experienced pilots did what they could to train others in more successful techniques, but without proper training exercises, these tactics were taught while in combat, resulting in additional unnecessary losses.

Without significant improvements to peacetime training exercises, the kill ratio fell to an average two to one and at one point falling below the one to one level during the Vietnam War, meaning the United States was losing more aircraft than the enemy forces were.

The USAF set forth an immediate study to resolve the problem, naming it "Red Baron" after the World War I German flying ace, Manfred Von Richthofen. The results of the study brought to light many issues that shouldn't have come as a surprise, but opened some eyes nonetheless.

Besides realizing that the flow of information about the enemy was insufficient, knowledge of their aircraft characteristics and level of training was essentially unknown. Because of this, USAF pilots were unable to train for effective aerial combat, and their aircraft were incapable of properly engaging the enemy with current load standards and flight characteristics. Most importantly, U.S. pilots found themselves being caught by surprise more often than not.

The USAF considered the current training programs sufficient for enabling pilots to enter combat and succeed. Results from the study showed rotating newer pilots into the combat role proved to be detrimental to their overall success rate. Although this took the USAF by surprise, there was little that could be done to fix the problem because pilots

Blasting off with the usual crowded ramp in the background, an Eglin Air Force Base–stationed F-15C Eagle from the 33rd Fighter Wing, 58th Fighter Squadron, heads out to the range. Aircraft from the 33rd in Florida and the 57th at Nellis typically work together because their roles both involve weapons testing and aircraft development. Red Flag exercises provide both squadrons with the ability to witness as well as try out new aerial tactics and maneuvers in a real-world combat environment.

were needed quickly and more advanced training wasn't possible at that time.

With the eventual conclusion of the Vietnam War in the 1970s, USAF leaders set forth a goal to revamp the training process and improve adversary tactics and familiarity with the enemy. The Red Baron study brought into focus the need not only to train pilots on a more personal level, but also to create an aircraft with the more specific role to achieve air superiority rather than using the multi-role F-4 Phantom. This change in direction led to the development of the highly agile McDonnell Douglas F-15 Eagle and eventually an entirely new breed of pilots.

Red Flag did not happen overnight. Many suggestions and concepts from numerous individuals slowly led to its creation. The January 1954 issue of *Fighter Gunnery* contained an article by Major Frederick C. "Boots" Blesse, Korean War fighter ace. In the article, Blesse said that positioning oneself at the proper angle constituted 85 percent of success in an air battle, while adjusting the pipper (symbol in an optical gun sight indicating where the weapon is aimed) was 10 percent, and the actual firing was only 5 percent. A change in Tactical Air Command (TAC) procedures suggesting training in dissimilar

Right: Major Derek "Tazz" Routt runs through a final preflight check prior to closing the canopy and beginning the long taxi to the other end of the airfield. Despite the proven need for an air force aggressor unit, a resulting squadron didn't come into being right away. Through years of bureaucracy and multiple levels of command, the 64th Aggressor Squadron was finally established in late 1972. Soon afterwards, the 65th AGRS and two overseas units were formed. Today, the 64th and 65th are a key part of one the most successful exercises in USAF history.

SCUE
1. PUSH BUTTON TO OP
2. PULL RING OUT 6 FE

F-15 Eagle

The F-15 is an all-weather, extremely maneuverable, tactical fighter designed to permit the air force to gain and maintain air supremacy over the battlefield.

The F-15's superior maneuverability and acceleration are achieved through its high engine thrust-to-weight ratio and low-wing loading. It was the first U.S. operational aircraft whose engines' thrust exceeded the plane's loaded weight, permitting it to accelerate even while in vertical climb.

The first F-15A flight was made in July 1972, and the first flight of the two-seat F-15B (formerly TF-15A) trainer was made in July 1973. In November 1974, the first Eagle was delivered to the 58th Tactical Fighter Training Wing at Luke Air Force Base, Arizona. In January 1976, the first F-15 destined for a combat squadron was delivered to the 1st Tactical Fighter Wing at Langley AFB, Virginia.

The single-seat F-15C and two-seat F-15D models entered the air force inventory beginning in 1979 and were first delivered to Kadena Air Base, Japan. These new models have Production Eagle Package (PEP 2000) improvements, including 2,000 pounds (900 kilograms) of additional internal fuel, provision for carrying exterior conformal fuel tanks and increased maximum takeoff weight of up to 68,000 pounds (30,600 kilograms).

The F-15E Strike Eagle is a dual-role fighter designed to perform air-to-air and air-to-ground missions. An array of avionics and electronics systems gives the F-15E the capability to fight at low altitude, day or night. The first production model of the F-15E was delivered to the 405th Tactical Training Wing, Luke AFB, Arizona in April 1988.

F-15C, D, and E models were deployed to the Persian Gulf in 1991 in support of Operation Desert Storm, where they proved their superior combat capability. F-15C fighters accounted for thirty-four of the thirty-seven air force air-to-air victories. F-15Es were operated mainly at night, hunting Scud missile launchers and artillery sites using the LANTIRN system.

They have since been used for air expeditionary force deployments and operations Southern Watch (no-fly zone in Southern Iraq), Provide Comfort in Turkey, Allied Force in Bosnia, Enduring Freedom in Afghanistan, and Iraqi Freedom.

Flying off the wing of a KC-135R tanker, F-15Cs from Langley AFB, Virginia, take turns refueling before heading back into combat over the Nellis Range Complex. The F-15 Eagle is one of the fastest fighter aircraft in the U.S. Air Force with a top speed in excess of 1,800 mph. A unique feature of the F-15 is its incredible maneuverability due to increased wing loading and a more powerful thrust-to-weight ratio. The larger wing area allows the aircraft to turn tighter with minimal loss in airspeed.

Replacing the initial squadron T-38s, the more agile F-5E Tiger II played the aggressor role for the 64th Aggressor Squadron, part of the 57th Fighter Weapons Wing. Shown here armed with two AIM-9J Sidewinder missiles, the single-seat F-5E proved to be a formidable adversary to Blue Team forces, allowing the USAF to design tactics around real-world enemy maneuvers. Combined with ground forces located throughout the Nellis Test and Training Range, Red Flag exercises were on their way to becoming the most intense warfighting simulations anywhere in the world. *USAF*

aircraft was the focus of an article published in the same magazine in March 1968.

The idea of dissimilar aircraft or the creation of an "aggressor" squadron floated around the air force for some time. Under the command of Major General R. G. "Zack" Taylor, the Tactical Fighter Weapons Center was created from the transformation of the 4520th Combat Crew Training Wing at Nellis AFB. Taking note of the enormous area surrounding Nellis, Taylor saw the range as providing great potential.

Colonel William L. Kirk at the Pentagon also noticed the need for more realistic training exercises. Working as part of the electronic combat directorate, Kirk, along with additional members of his staff in Washington, D.C., began brainstorming for new ways to improve training. Of note was the Foreign Technology Division at Wright-Patterson AFB in Ohio. The division operated a number of Soviet aircraft that could be utilized to provide a realistic combat environment based on their unique maneuverability and air combat tactics. Unfortunately, the project was soon dismissed due to administrative difficulties.

Dissatisfied with the loss rate during the Vietnam War, Air Force Chief of Staff General John D. Ryan approved a proposal made by Kirk and Major John A. Corder. The proposal consisted

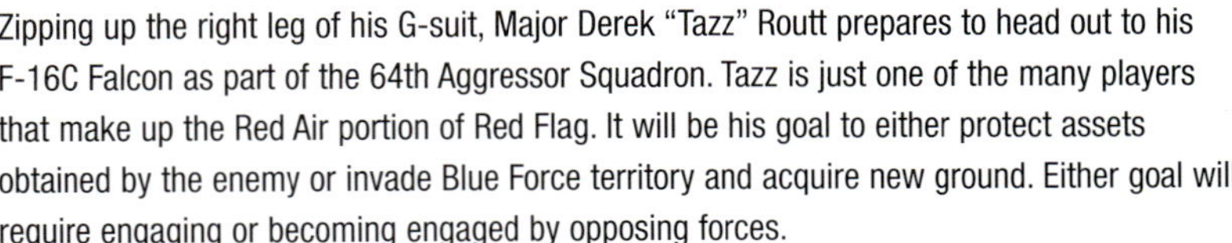
Zipping up the right leg of his G-suit, Major Derek "Tazz" Routt prepares to head out to his F-16C Falcon as part of the 64th Aggressor Squadron. Tazz is just one of the many players that make up the Red Air portion of Red Flag. It will be his goal to either protect assets obtained by the enemy or invade Blue Force territory and acquire new ground. Either goal will require engaging or becoming engaged by opposing forces.

of creating an air-to-air aggressor squadron that would utilize surplus Northrop T-38 Talons as was suggested by Lieutenant Colonel Charles A. Homer. The squadron would visit other units in the field as well as have those units circulate through Nellis to train with the aggressors at the nearby Nellis Range Complex.

It wasn't until fall of 1972 that the air force established the 64th Fighter Weapons Squadron. Simulating MiG-21s, the squadron initially flew T-38s and later added F-5E Tiger II aircraft. Exercises utilizing these small, agile aircraft were deemed so successful that the USAF established the 65th Fighter Weapons Squadron as Nellis' second aggressor unit.

Even with the creation of the two aggressor squadrons as well as two additional overseas units, Major Richard Moody Suter had bigger plans. Realizing the potential for large-scale combat-training operations as well as what the vast Nellis Range had to offer, he intended to go beyond basic air-to-air combat maneuvering. Suter expanded on Homer's and Corder's original ideas with the aggressor

A lone F-16CG from Hill Air Force Base's 388th Fighter Wing, 421st Fighter Squadron, the Black Widows, cruises over the Nellis Range Complex. Training fighter pilots has always proven to be a daunting task. The most difficult proposition facing military officials was how to make training as realistic as possible without incurring unnecessary losses. Even with the creation of four aggressor squadrons, individuals like Major Richard Moody Suter and General Robert J. Dixon forged ahead to create an entirely new exercise—Red Flag.

squadrons and laid out his concept of what would eventually become Red Flag.

Suter, a charismatic individual, was once described as a man who performed systems management before systems management was invented. His concept of what future training exercises could be didn't come without a myriad of challenges. Aircraft losses were on the decline and the USAF was content with the falling numbers. In 1951, losses were at 824 aircraft. In 1959, the number dropped to 472, and in 1965, the number of lost aircraft declined to 262. Safety had become of such paramount importance that properly training aviators in the role of combat maneuvering had become improbable. Suter's challenge was to find a way to conduct realistic training exercises while adhering to the U.S. Air Force's desire to maintain flying safety.

Allied Aircraft: GR4 Toronado

The Tornado originated from a late-1960s feasibility study into multi-role combat aircraft conducted by Belgium, Canada, Germany, Italy, the Netherlands, and the UK. When Belgium, Canada, and the Netherlands later withdrew, a new tri-national company, Panavia, was set up to build the aircraft, with the work divided among Italy, Germany, and the UK.

The Tornado was ultimately selected to meet the UK requirement for an air defense fighter in 1971. The first prototype, which was German-built, flew on August 14, 1974. The final prototype selected for production, the GR1, first flew in 1979. The initial RAF requirement was for 220 aircraft, and the first of these was delivered to the new Tri-national tornado training establishment (TTTE) at RAF Cottesmore in July 1980. The first frontline squadron to reequip with Tornado was IX (B) Squadron at Honington from June 1982 and now stationed at RAF Marham. The RAF received the first F2 variant in 1984, the same year the UK Ministry of Defence began studying the G4 upgrade. In 1985, the UK signed a contract with Saudi Arabia for the sale of 72 Tornado models, with the first delivered the following year. The F3 entered UK service in 1986, first seeing action in the 1991 Gulf War.

The GR4 upgrade wasn't approved until 1994; the first flight of the GR4 came three years later. Later in 1997, the RAF received their first GR4s, which entered frontline service in 1998. Deliveries of the aircraft were completed in 2002, and in 2003 Tornado GR4s figured prominently in the Iraq War. The final new-build Tornado was in 1998 for Saudi Arabia, with all work now focusing on the various upgrades being conducted.

A dedicated reconnaissance version, the GR4A, is also in RAF service. Many GR1s and GR1Bs are undergoing a mid-life update program, and this will see updates to many internal systems and defensive aids and extend the service life of the Tornado for some fifteen years or so, until replaced by the future offensive air system (FOAS) due in 2015. These updated aircraft are designated GR4s.

Calling RAF Marham its home, this GR4 Tornado from No. IX Squadron blasts down runway 3 left heading towards the Nellis Range Complex. One of the largest allied supporters of Red Flag exercises, the UK routinely sends crews to Nellis AFB. It's not uncommon to see a row of GR4 Tornados and GR3 Jaguars as well as a reconnaissance R1 Nimrod and support tanker VC10 on the Nellis ramp during Red Flag. The No. IX Squadron has been operating the GR4 Tornado since 1982.

With a Massachusetts-based A-10 Thunderbolt II in the foreground, rows and rows of military hardware rest on the Nellis ramp. It is not uncommon to see upwards of 150 aircraft from all over the world lined up underneath the Las Vegas skyline. In the distance, world-renowned landmarks such as the Stratosphere, Rio, Wynn, and other tempting spots beckon the visiting Red Flag participants. Though perhaps a blessing in disguise, the rigorous schedules limit the amount of time one has to play in Sin City.

Studies had shown that most losses were incurred during the pilot's first ten combat missions. Suter knew of these studies and argued that a few losses during training would prevent massive losses in the field. At this point, exercises were very standardized, resulting in an extreme lack of training but a satisfactorily low number of incidents. Suter showed the air force that after ten combat missions, losses dropped to nearly zero and proposed creating a realistic training environment so accurate that a pilot could log his first ten missions. This controlled environment would teach pilots how to act quickly, precisely showing them the consequences of their mistakes without threatening their life or career. It was hoped that by entering combat after these training exercises both pilots and the USAF brass would have more confidence knowing they had survived their most vulnerable period.

Once Suter had received the various approvals from the Pentagon, he proceeded to brief General Robert J. Dixon, commander of the Tactical Air Command. Suter's briefing tackled the future of air combat at a much higher level of sophistication than was currently employed. Training tactics of the day were extremely predictable and routine, a basic evolution of World War II fundamentals. Suter illustrated that a successful exercise needed to involve a program that provided realistic training against a realistic threat in order to properly test hardware and air-combat tactics. He suggested that Nellis Air Force Base not only hold Red Flag exercises as a proving

A/OA-10 Thunderbolt II

The A/OA-10 Thunderbolt II is the first air force aircraft specially designed for close air support of ground forces. They are simple, effective, and survivable twin-engine jet aircraft that can be used against all ground targets, including tanks and other armored vehicles.

The first production A-10A was delivered to Davis-Monthan Air Force Base, Arizona, in October 1975. It was designed specially for the close air support mission and had the ability to combine large military loads, long loiter, and wide combat radius, which proved to be vital assets to the United States and its allies during Operation Desert Storm and Operation Noble Anvil in Kosovo.

In the Gulf War, A-10s had a mission-capable rate of 95.7 percent, flew 8,100 sorties and launched 90 percent of the AGM-65 Maverick missiles.

Built around a massive 30mm GAU-8/A seven-barrel Gatling gun, the A-10 Thunderbolt II is more commonly referred to as the Warthog for its down-and-dirty looks. With survivability in mind, the A-10 was overdesigned with a redundant hydraulic system backed up by a third manual system and components that could be interchanged regardless of which side they were to be used on. The two TF34-GE-100 turbofan engines were placed high and shielded by the aircraft's elevator. The most unique aspect of the A-10 is the fact that the pilot sits in a tub of titanium.

ground for training pilots but also as a laboratory where the air force could effectively test possible solutions to new challenges they discovered in training.

Suter visited squadrons all over the world spreading word about the Red Flag program and obtaining support for the exercise. He explained that Red Flag would hone the skills of the pilot not only to save lives but also to make them more effective in combat. Red Flag would improve precision bombing and efficient aerial dominance resulting in a greater number of victories on the battlefield.

Suter had Dixon's undivided attention and the idea was quickly approved. Dixon's operational deputy, Major General Charles A. Gabriel, and Major General James A. Knight, commander of the Tactical Fighter Weapons Center, were told to establish Red Flag within six months. Colonel Richard Murray, Dixon's comptroller, was tasked with finding the money to make it all happen.

Returning home after sunset, two A-10 Thunderbolt IIs from the 22nd Test and Evaluation Squadron (TES) line up to land on runway 27L. The 422nd TES conducts operational tests for Air Combat Command (ACC) on new hardware and upgrades to each of five different unit aircraft in a simulated combat environment. Along with the A-10, other aircraft belonging to the 422nd TES include the F-15C Eagle, the F-15E Strike Eagle, the F-16C Fighting Falcon and the HH-60G Pave Hawk helicopter.

Dixon seemed the perfect individual to push the Red Flag concept to approval. Along with the shared interest of Air Force Chief of Staff General David C. Jones, the two agreed to try out the idea of realistic training so long as Tactical Air Command could keep the accident rate below seven per one hundred thousand flight hours.

With the program approved, Suter went right to work. Enlisting the assistance of Colonel P. J. White, Colonel David Burney, Lieutenant Colonel Marty Mahrt and civilian computer expert Ned Greenhalgh, the small staff went to work establishing the groundbreaking training program. The program was complicated by the desire to include every aspect of warfare including bombers, fighters, tankers, electronic countermeasures (ECM), reconnaissance aircraft, and more. The enemy would be equipped with antiaircraft and integrated missile systems, advanced radar systems, and, of course, highly effective dissimilar aerial interceptors. With 232 Vietnam combat missions under his belt, Suter was well respected in the Nellis community, and many fighter pilots supported his ideals. Little did anyone know at the time, Red Flag would become one of the most successful and instrumental training programs in aviation history.

Just as Dixon requested, the first Red Flag was held exactly on schedule. On November 29, 1975, thirty-seven aircraft supported by 561 people flew 552 sorties over the Nevada skies. Despite concentrating mainly on air-to-surface training, the exercise still demonstrated a large amount of air-to-air combat. Though far from today's numbers, the first Red Flag was an unqualified success.

Due to eight aircraft being lost during the first two years, air force commanders initially scrutinized Dixon. Their concern was creating the illusion of poor leadership. Fortunately, Dixon maintained his course and the accident rate began to fall to a point lower than that of the air force loss rate as a whole. When the air

HL
HL
AF
88
512
SN1735

force began taking note of the exercise's success, the USAF Systems Command wanted to use Red Flag for operational testing and evaluation. Dixon chose instead to let the major commands contribute ideas for the further development of Red Flag toward a more intense training platform.

Red Flag inspired numerous additional exercises that would encompass almost every aspect of air combat. Over time, other services would join in with units from around the world. Established at Hurlburt Field, Florida, Blue Flag was designed with the European Theater in mind and the diverse technology and conditions unique to that part of the world. Green Flag worked with Red Flag exercises as a means of integrating electronic countermeasure warfare. Maple Flag was developed to train Canadian air forces in a similar manner to Red Flag.

The various flag exercises had become so successful, not to mention intense, that training standards had become more rigorous than actual combat. When Dixon had completed his tour at the Tactical Air Command, Red Flag exercises were handed over to General W. L. Creech, who further improved and accelerated the Red Flag exercises to what they are today. Today's Red Flag encompasses 250 different units and approximately 750 aircraft of all types annually. More than twelve thousand sorties and twenty-one thousand flight hours will be accumulated with the support of over eleven thousand crewmembers.

As retired air force colonel and author Walter J. Boyne noted, a U.S. Air Force pilot returning from a combat mission over Iraq during the 1991 Gulf War was heard to remark, "It was almost as intense as Red Flag."

In early 2005, the United States Department of Defense assembled the largest exercise of the fiscal year. With a cost of $21 million and participation by more than ten thousand personnel at forty-four different locations, Joint Red Flag (JRF) was intended to take current exercises to the next level. So complex was JRF that only one exercise has been performed.

Two F-16Cs from Hill Air Force Base fly tight formation during a shallow banking turn. Noticeable on the nearest pylon is a long dart-like probe called the Nellis Air Combat Training System (NACTS). The NACTS pod can track up to one hundred individual aircraft denoting their altitude, air speed, range positioning, and more for a complete assessment of actions to be reviewed in detail at the mass debrief following the mission.

Chapter 3

THE RED TEAM

Pilots of the 64th Aggressor Squadron debrief within the confines of their Eastern Bloc environment. Aggressor pilots are trained not necessarily to win, but rather provide qualified training to members of the Blue Team. At the risk of harming the morale of the newer pilots, Red Air establishes a crawl, walk, run approach to gradually introduce Blue Air pilots to the tactics of various levels of hostile aircraft.

An aspect of aerial combat training unique to Nellis Air Force Base Red Flag exercises would undoubtedly be the Red Team. Walking through the corridors of the 64th Fighter Weapons Squadron (FWS), one would think they had been relocated to the cold-war Soviet Union. Images of Stalin accompanied by illustrations of the hammer and sickle adorn the walls with patches of crimson red. With the exception of the acronym-laden pilot lingo, deep down inside, they are still fortunately on the good side.

The 64th FWS came as the result of years of bureaucracy and lessons learned from failed combat training tactics. As the U.S. Air Force watched the kill ratio drop to an unprecedented level, aerial combat training in the states concentrated on safe flying and minimal risks. Pilots developed a routine of knowing the scenario ahead of time by flying repetitive training sorties. This resulted in enemy aircraft taking U.S. forces by surprise and introducing aerial maneuvers and tactics unfamiliar to American pilots.

To further add to the problems American pilots faced, combat sorties typically incorporated multi-role missions involving mainly air-to-surface bombing tasks. This made combat maneuvering nearly impossible without prematurely

High over the Nellis Range Complex, an F-4G Phantom II Wild Weasel belonging to the 563rd Tactical Fighter Squadron, 35th Tactical Fighter Wing, banks to the right with an AGM-45 Shrike missile on the right wing. In October 1978, the 563rd received new aircraft from Nellis Air Force Base, Nevada, to become the first operational squadron to fly the advanced F-4G Wild Weasel. *USAF*

With the proven theory of having an aggressor squadron flying dissimilar aircraft, the 57th Fighter Wing, consisting of the 64th and 65th Aggressor Squadrons, proudly flew the nimble F-5E Tiger II. Seen here flying in wedge formation over nearby Lake Mead, Nevada, in 1981, tactics went beyond simply moving to different aircraft and began incorporating various camouflage paint schemes. *USAF*

disposing of ordnance. Enemy forces were winning on both fronts.

Back home, pilots were in short supply and efficient rotation of new pilots to the battlefront was seen as the key to success. Proper training simply wasn't an option. Learning combat tactics consisted of flying against similar aircraft such as the bulky F-4 Phantom. At times, aircraft from the same squadron would fly against each other under predetermined and controlled conditions. This type of training couldn't be further from what was being faced in an actual combat environment.

With the development of the improved F-4E, Colonel John Boyd put into practice revised maneuverability concepts. However, even with these improvements, these new tactics still failed to address the problem.

Established on March 3, 1969, the U.S. Navy developed a school to train fleet fighter pilots in air combat tactics to counter the relatively poor air-combat performance being experienced by navy aircrews over Vietnam. Officially known as the United States Navy Fighter Weapons School, Top Gun instructors utilized Douglas A-4 Skyhawk IIs and F-5s to train navy F-4 aircrews. The use of dissimilar and more agile aircraft to teach aerial combat tactics to crews flying the larger Phantoms was a huge success for the navy.

Seeing the benefits of dissimilar aircraft training and utilizing the lessons learned from the Red Baron study, Colonel William L. Kirk working together with Major John A. Corder

F-16 Fighting Falcon

The F-16 Fighting Falcon is a compact multi-role fighter aircraft. It is highly maneuverable and has proven itself in air-to-air combat and air-to-surface attack. It provides a relatively low-cost, high-performance weapon system for the United States and allied nations.

The F-16A, a single-seat model, first flew in December 1976. The first operational F-16A was delivered in January 1979 to the 388th Tactical Fighter Wing at Hill Air Force Base, Utah.

The F-16B, a two-seat model, has tandem cockpits that are about the same size as the one in the A model. Its bubble canopy extends to cover the second cockpit. To make room for the second cockpit, the forward fuselage fuel tank and avionics growth space were reduced. During training, the forward cockpit is used by a student pilot with an instructor pilot in the rear cockpit.

All F-16s delivered since November 1981 have built-in structural and wiring provisions and systems architecture that permit expansion of the multi-role flexibility to perform precision strike, night attack, and beyond-visual-range interception missions. This improvement program led to the F-16C and F-16D aircraft, which are the single- and two-place counterparts to the F-16A/B, and incorporate the latest cockpit control and display technology. All active units and many Air National Guard and Air Force Reserve units have converted to the F-16C/D.

The F-16 was built under an unusual agreement creating a consortium between the United States and four NATO countries: Belgium, Denmark, the Netherlands, and Norway. These countries jointly produced with the United States an initial 348 F-16s for their air forces. Final airframe assembly lines were located in Belgium and the Netherlands. The consortium's F-16s are assembled from components manufactured in all five countries. Belgium also provides final assembly of the F100 engine used in the European F-16s. Recently, Portugal joined the consortium. The long-term benefits of this program will be technology transfer among the nations producing the F-16, and a common-use aircraft for NATO nations. This program increases the supply and availability of repair parts in Europe and improves the F-16's combat readiness.

USAF F-16 multi-role fighters were deployed to the Persian Gulf in 1991 in support of Operation Desert Storm, where more sorties were flown than with any other aircraft. These fighters were used to attack airfields, military production facilities, Scud missile sites, and a variety of other targets.

During Operation Allied Force, USAF F-16 multi-role fighters flew a variety of missions to include suppression of enemy air defense, offensive counter air, defensive counter air, close air support, and forward air controller missions. Mission results were outstanding as these fighters destroyed radar sites, vehicles, tanks, MiGs, and buildings.

Since September 11, 2001, the F-16 has been a major component of the combat forces committed to the Global War on Terror flying thousands of sorties in support of operations Noble Eagle (Homeland Defense), Enduring Freedom in Afghanistan, and Iraqi Freedom.

An F-16CJ from Cannon Air Force Base's 27th Fighter Wing, 522nd Fighter Squadron, shows off the sleek lines of the Fighting Falcon, commonly referred to by pilots as the Viper. Of note is the Link 16 SNIPER XR advanced targeting pod (ATP) located just below the intake, as well as the joint helmet-mounted cueing system (JHMCS) worn by the pilot. The Sniper XR pod incorporates a high-resolution, mid-wave, third-generation forward-looking infrared (FLIR), a dual-mode laser, and a CCD-TV, along with a laser spot tracker, a laser marker, and a sophisticated data-link.

Headquarters of the 64th and 65th Aggressor Squadrons. Besides the 26th and 527th Space Aggressor Squadrons, the 64th and 65th are the only units designated to portray the enemy and thus display a unique camouflage paint scheme. Operating under the 414th Combat Training Squadron (CTS), the aggressor units were originally designed to visit other bases and train where needed. Because of the Nellis Range Complex, it was deemed more suitable for the various units to visit the aggressors. Only recently have the 64th and 65th taken their mission on the road by participating in Red Flag—Alaska out of Eielson AFB.

took note of the aircraft operated by the Foreign Technology Division at Wright-Patterson Air Force Base. Use of these Soviet aircraft could prove invaluable to future training efforts. Unfortunately, due to administrative difficulties, the acquisition of these aircraft proved impossible.

Sharing his dissatisfaction with the loss rate incurred during the Vietnam War, Air Force Chief of Staff General John D. Ryan approved a proposal made by Kirk and Corder to assemble an air-to-air aggressor squadron. Although the 64th Pursuit Squadron was activated in 1941, they were designated the 64th Fighter Weapons Squadron and reestablished on September 7, 1972, operating under the 414th Combat Training Squadron.

Utilizing surplus T-38 Talons, the improved training techniques were deemed so successful that the air force commissioned the 65th FWS in 1975 as well as two additional overseas units for Pacific and European Theater exercises. The first assigned aircraft for the 65th FWS was the F-5E Tiger II. The 64th moved over to the F-5E at the same time.

In August 1975, the 26th Aggressor Squadron (26th TFTS) was activated at Clark Air Base, the Philippines, to provide dissimilar air-to-air combat tactics (DACT) to Pacific Air Force fighter forces. On January 1, 1976, the 52nd Tactical Fighter Training Aggressor Squadron (527th TFTS) was activated at RAF Alconbury, United Kingdom, to provide DACT to United States air forces in Europe. On April 1, 1989, the 64th FWS transitioned from F-5Es to General Dynamics F-16As. Due to budget constraints, the 65th FWS, 527th TFTS, and 26th TFTS were deactivated in late 1989. In October 1989, the 64th FWS converted to F-16Cs. The 65th FWS was reestablished in January 2006 flying the F-15C Eagle.

With refueling door open, this F-16C from the 64th Aggressor Squadron prepares to take on fuel from an awaiting KC-135 Stratotanker. Due to advancements in radar and electronics as well as the increased range, the 64th chose to transition to the General Dynamics F-16A in 1989 leaving behind the agile F-5Es. The blue-gray camouflage found on this aircraft as well as the other schemes throughout the 64th are unique to the squadron.

Both the 64th and 65th operate under the 414th Combat Training Squadron. The squadron was originally designated the 414th Night Fighter Squadron on January 21, 1943, at Orlando Army Air Base in Florida. The 414th NFS was assigned to the AAF School of Applied Tactics flying the Douglas A-20 attack bomber and P-70 night fighter and British Bristol Beaufighter.

The squadron was reassigned to the Twelfth Air Force on May 10, 1943, followed by the 2nd Air Defense (later, 63rd Fighter) Wing on May 29, 1943. It was then assigned to the 62nd Fighter Wing in 1944 and then to the XXII Tactical Air Command on April 1, 1945. And later, back to the Twelfth Air Force, on June 7, 1945. The squadron saw combat in Mediterranean and European theater operations from 1943 to 1945 operating from locations in Corsica, Algeria, Tunisia, Sardinia, Italy, Belgium, and Germany. The various aircraft flown by the unit included the Lockheed P-38 Lightning (1945); the North American P-51 Mustang (1945, 1946–1947); and the Northrop P-61 Black Widow (1945, 1946–1947).

With a hiatus in operational duties from June 1945 to August 1946, the squadron was reassigned to the Fourth Air Force and returned to the United States at Lemoore Army Airfield in California. It was reassigned to Air Defense Command on March 21, 1946, and then to the Tactical Air Command in July 1946. Next, the squadron was sent to the

E-3 Sentry

The E-3 Sentry is an airborne warning and control system (AWACS) aircraft that provides all-weather surveillance, command, control, and communications needed by commanders of U.S., NATO, and other allied air defense forces.

Engineering, testing, and evaluation began on the first E-3 Sentry in October 1975. In March 1977, the 552nd Airborne Warning and Control Wing (now 552nd Air Control Wing, Tinker Air Force Base, Oklahoma) received the first E-3s.

NATO has acquired seventeen E-3As and support equipment. The first E-3 was delivered to NATO in January 1982. The United Kingdom has seven E-3s, France has four, and Saudi Arabia has five.

As proven in operations Desert Storm, Allied Force, Enduring Freedom, and Iraqi Freedom, the E-3 Sentry is the premier air battle command-and-control aircraft in the world. AWACS aircraft and crews were instrumental to the successful completion of operations Northern and Southern Watch, and are still engaged in operations Noble Eagle and Enduring Freedom. They provide radar surveillance and control in addition to providing senior leadership with time-critical information on the actions of enemy forces.

The data collection capability of the E-3 radar and computer subsystems allowed an entire air war to be recorded for the first time in the history of aerial combat.

During the spring of 1999, the first AWACS aircraft went through the radar system improvement program (RSIP). RSIP is a joint U.S./NATO development program that involved a major hardware and software intensive modification to the existing radar system. Installation of RSIP enhanced the operational capability of the E-3 radar electronic countermeasures, and improved the system's reliability, maintainability, and availability.

Based out of Tinker AFB in Oklahoma, this E-3 Sentry heads North towards the Nellis Range Complex. The dome is thirty feet in diameter, six feet thick, and eleven feet above the fuselage. It contains a radar subsystem that permits surveillance from the Earth's surface up into the stratosphere, over land or water, with a range of over two hundred miles.

Third Air Force in October 1946 and then to the Ninth Air Force in November. It was then moved to the 6th Fighter Wing in March 1947. The much-traveled squadron was then inactivated on September 1, 1947, at the Rio Hato in Panama.

As the makings for Red Flag exercises and the need for aggressor squadrons neared, the dormant 414th NFS was redesignated as the 414th Fighter Weapons Squadron on August 22, 1969. The squadron was reactivated in October at Nellis AFB, Nevada. Assigned to the 57th Fighter Wing, the 414th was tasked with combat crew training using F-4 Phantom II aircraft. The squadron was again deactivated on December 30, 1981.

The 414th FWS was redesignated as the 414th Composite Training Squadron and reactivated nearly ten years later, on November 1, 1991, at Nellis AFB. Assigned to the 57th Wing's Operations Group and flying the F-16, the squadron was tasked with conducting Red Flag exercises. The 414th was subsequently redesignated as the 414th Training Squadron in January 1993 and finally as the 414th Combat Training Squadron on July 1, 1994.

The two aggressor squadrons based at Nellis AFB fly the F-16 and F-15 fighters, the F-16Cs belonging to the 64th Aggressor Squadron and the F-15Cs belonging to the 65th.

Touching down on runway 21-Right after a morning mission, this freshly painted McDonnell Douglas F-15C Eagle is the first of twenty-two Eagles destined for the newly reestablished 65th Aggressor Squadron. Just as in the past, the 64th and 65th will once again work together as aggressor units training Blue Four on the intricacies of combat tactics. Similar to the 64th, the Eagles will also carry a variation of camouflage paint schemes ranging from brown to blue.

The F-16 employs advanced aerospace science and proven reliable systems from other aircraft such as the F-15 and General Dynamics F-111. These were combined to simplify the airplane and reduce its size, purchase price, maintenance costs, and weight, making it a formidable replacement for the F-5E fighters that preceded them. The light weight of the F-16 fuselage is achieved without reducing its strength. With a full load of internal fuel, the F-16 can withstand up to nine G's (nine times the force of gravity), which exceeds the capability of other current fighter aircraft. These increased tolerances, along with improved agility, allow Red Team pilots (often called Red Air) the ability to replicate or exceed flight characteristics of the hostile aircraft they replicate.

The cockpit and its bubble canopy give the pilot unobstructed forward and upward vision along with greatly improved vision over the side and to the rear. The seat-back angle was expanded from the usual 13 degrees to 30 degrees, increasing pilot comfort and gravity-force tolerance. The pilot has excellent flight control of the F-16 through its fly-by-wire system. Electrical wires relay flying commands, replacing the usual cables

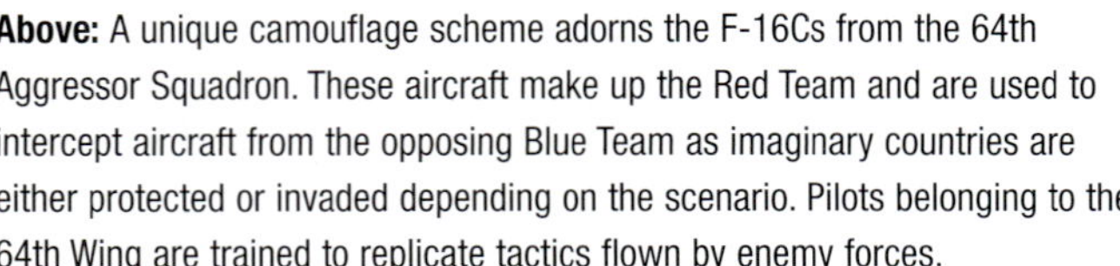

Above: A unique camouflage scheme adorns the F-16Cs from the 64th Aggressor Squadron. These aircraft make up the Red Team and are used to intercept aircraft from the opposing Blue Team as imaginary countries are either protected or invaded depending on the scenario. Pilots belonging to the 64th Wing are trained to replicate tactics flown by enemy forces.

Right: Reestablished on January 12, 2006, this F-15C Eagle wears the typical camouflage of 64th Aggressor Squadron F-16s. Going by the call sign "MiG-7," this is the first of twenty-two F-15s expected to take on the aggressor role of fighters from the 65th. Paint schemes will resemble those of the 64th, which have blue-gray and desert-tan camouflage.

and linkage controls. For easy and accurate control of the aircraft during high-G-force combat maneuvers, a side-stick controller is used instead of the conventional center-mounted stick. Hand pressure on the side-stick controller sends electrical signals to actuators of flight-control surfaces such as ailerons and rudder.

Maintaining the desire to participate and train with dissimilar aircraft, the F-15C was employed recently to accompany the F-16 force on Red Air. The F-15C Eagle's air superiority is achieved through a mixture of unprecedented maneuverability and

acceleration, range, weapons, and avionics. It can penetrate enemy defense and outperform and outfight any current enemy aircraft. The F-15 has electronic systems and weaponry to detect, acquire, track, and attack enemy aircraft while operating in friendly or enemy-controlled airspace. The weapons and flight-control systems are designed so one person can safely and effectively perform air-to-air combat.

The F-15's superior maneuverability and acceleration are achieved through high engine thrust-to-weight ratio and low wing loading. Low wing loading (the ratio of aircraft weight to its wing area) is a vital factor in maneuverability and, combined with the high thrust-to-weight ratio enables the aircraft to turn tightly without losing airspeed.

The Team

Without a doubt, it is the Red Team that makes the Red Flag experience so successful. With training centered around combat tactics of various forces and opposing aircraft, pilots of the Red

An image taken in the early 1990s shows pilots at a Red Flag mission debriefing using what was once a state-of-the-art mission debriefing system. Each session provided instant feedback to participants in tactical warfare training. Designed by members of the 4440th Tactical Fighter Training Group, the Nellis Test and Training Range introduced pilots to obstacles such as those encountered by fighter aircraft during Operation Desert Storm. Besides advancements in debriefing technology and information acquisition, little has changed in the objective of Red Flag. *USAF*

Team alter their maneuvers to relate as closely as possible to the combatants they are most likely to face in actual combat. The goal of the Red Team is not to win or lose, but to teach.

In a typical Red Flag exercise, friendly Blue Forces engage hostile Red Forces in combat situations. Blue Forces are made up of units from ACC, Air Mobility Command (AMC), U.S. Air Force in Europe, Pacific Air Forces (PACAF), Air National Guard (ANG), United States Air Force Reserves (USAFR), U.S. Army, Navy, Marine Corps and allied air forces. They are led by a Blue Forces commander who orchestrates the deployment plan. Red Forces are composed of Red Flag's Adversary Tactics Division flying the F-16 and providing air threats by emulating enemy tactics. They are often augmented by other U.S. Air Force, Navy, and Marine Corps units flying in concert with electronic ground defenses and communications and radar jamming equipment.

Flying camouflaged F-16s and F-15s, the heart of Red Air is to provide Blue Four with their first experience participating in combat scenarios. Blue Four is named for lieutenants and captains competent in their aircraft but without flying experience in a composite strike force along with being the outermost pilot in a four-ship echelon formation. Their training, along with the rest of the aerial expeditionary force will be conducted in steps of increasing difficulty as the two-week session progresses.

Major Andy "Popeye" Hansen has been behind the stick of an air force jet for thirteen years and, as of 2006, he's been flying

F-117A Nighthawk

The F-117A Nighthawk is the world's first operational aircraft designed to exploit low-observable stealth technology. This precision-strike aircraft penetrates high-threat airspace and uses laser-guided weapons against critical targets.

The F-117A production decision was made in 1978 with a contract awarded to Lockheed Advanced Development Projects, the "Skunk Works," in Burbank, California. The first flight over the Nevada test ranges was on June 18, 1981, only thirty-one months after the full-scale development decision.

Streamlined management by Aeronautical Systems Center, Wright-Patterson Air Force Base, Ohio, combined breakthrough stealth technology with concurrent development and production to rapidly field the aircraft.

The first F-117A was delivered in 1982, and the last delivery was in the summer of 1990. Air Combat Command's only F-117A unit, the 4450th Tactical Group (now the 49th Fighter Wing, Holloman Air Force Base, New Mexico), achieved operational capability in October 1983.

During Operation Desert Storm in 1991, F-117As flew approximately 1,300 sorties and scored direct hits on 1,600 high-value targets in Iraq. It was the only U.S. or coalition aircraft to strike targets in downtown Baghdad. Since moving to Holloman AFB in 1992, the F-117A and the men and women of the 49th Fighter Wing have deployed to Southwest Asia more than once. On their first trip, the F-117s flew nonstop from Holloman to Kuwait, a flight of approximately 18.5 hours—a record for single-seat fighters that stands today.

In 1999, 24 F-117As deployed to Aviano Air Base, Italy, and Spangdahlem Air Base, Germany, to support NATO's Operation Allied Force. The aircraft led the first allied air strike against Yugoslavia on March 24, 1999.

Returning to the skies over Baghdad, F-117As launched Operation Iraqi Freedom with a decapitation strike on March 20, 2003. Striking key targets in the toppling of Saddam Hussein's regime, twelve deployed F-117s flew more than one hundred combat sorties in support of the global war on terrorism.

The F-117A program demonstrates that stealth aircraft can be designed for reliability and maintainability. It created a revolution in military warfare by incorporating low-observable technology into operational aircraft. The aircraft receives support through a Lockheed-Martin contract known as total system performance responsibility.

Utilizing sharply angled faceted edges, the F-117A Nighthawk introduced the world to the reality of a new kind of technology—stealth. Stationed at Holloman Air Force Base, New Mexico, these aircraft remained a mystery until their unveiling in the late 1980s. Today the technology brought to light by the F-117 is becoming outdated, and faceted edges are no longer needed to obtain stealth characteristics. Rounded, swooping lines, as illustrated by the B-2A Spirit bomber and F-22 Raptor combined with applied radar absorbent materials (RAM), are allowing aircraft to be faster and more efficient.

Colonel Michael "Muff" McKinney takes on the call sign "MiG-3" during a morning sortie. Flying an F-16C Viper from the 64th Aggressor Squadron, Colonel McKinney awaits approval from ground control to begin his taxi to the south end of Nellis AFB. From there he will launch from runway 3L, head out to the Nellis Range Complex, and wreak havoc with Blue Air pilots aiming to advance into enemy airspace or defend their given territory.

with the 64th Aggressor Squadron for two years. "We work in a building-block approach to gradually enhance a pilot's reaction level, and throughout the course of the program events are tailored to provide an increasing challenge."

Pilots of the 64th and the newly reestablished 65th Aggressor Squadrons are welcomed to the units through a series of steps. Initially each new pilot is given around nine to twelve syllabus rides and orientation flights to help gear up to their "check qualifications." These flights include a detailed tour of the Nellis Range Complex and familiarization with no-fly zones as well as key targeting areas vital to future Red Air and Blue Air tactics. Based on the mission flown, each Red Air pilot is given a call sign, with "MiG-1" being the instructor.

Before each mission, the Red Team will gather for a planning conference where they will discuss the various steps and scenarios available to them based on the current training levels of Blue Air. During the planning conference, Red Air pilots will adjust tactics, levels of defensive reactions, and weapons selection unique to the current mission. Even the type of aircraft and the properties associated with its maneuverability can be implemented.

Once in the air, the Red Team typically stakes claim to the northwest region of the Nellis Range Complex. Their goal is to either defend a make-believe country from the advances of the Blue Team or infiltrate the Blue Team's imaginary country to the southeast. Unlike the Blue Team, Red Air has the ability to regenerate if "shot down" during engagement.

At the 64th Aggressor Squadron headquarters, Captain Bryan "Groucho" Duke takes on the MiG-1 position coordinating the prebrief with Kool, Muff, Lamont, Bolt, and Tazz, rounding out Red Air for this particular sortie. Qualified aggressor pilots will take turns assuming the role of MiG-1, the lead aircraft of Red Air. Prebrief discussions include the mission objective, weather conditions, no-fly zones, tactic intensity, as well as the usual preflight paperwork.

But the Red Team goes beyond aerial combat. Scattered throughout the Nellis Test and Training Range are mobile missile systems, antiaircraft guns, surface-to-air missiles, and radar installations. Obviously these weapons are not actually launched or fired, but rather they track and target Blue Force aircraft electronically adding to the level of difficulty each pilot must face.

Upon completion of each mission, everyone gathers at Suter Hall for a mass debriefing sometimes lasting a couple of hours. Data is retrieved from the Nellis Air Training System (NATS) instrumentation pods that capture and transmit data about position, altitude, speed, maneuvers, and weapons employment. This data is invaluable for debriefings because it enables accurate assessments of engagements for up to one hundred aircraft at a time.

Regardless of the mission, it is Red Air's job to make the lives of the Blue Force difficult. Because real-world air combat scenarios are unpredictable and pilots face the possibility of extreme overtasking, it is the job of Red Air to see that U.S. Air Force pilots don't face that situation for the first time over hostile territory. Situations involving defensive and offensive tactics as well as search and rescue are all covered within the pilots' first ten missions at Red Flag, and Red Air is there to ensure that Blue Four pilots learn all they can before actual deployment of the aerial expeditionary forces (AEF) into combat.

Predictions based on the potential success of an aggressor squadron called for plans to make Red Air a "road show." Instead of having an entire AEF travel to Nellis, the two squadrons would make appearances at various bases to provide training where necessary. Due to the immense training facility available to forces just outside Las Vegas, it was deemed practical for AEF forces to travel to Nellis instead. With the recent change of Cope Thunder to Red Flag–Alaska, the prospects of a traveling

KC-10 Extender

The KC-10 Extender is an Air Mobility Command advanced tanker and cargo aircraft designed to provide increased global mobility for U.S. armed forces. Although the KC-10's primary mission is aerial refueling, it can combine the tasks of a tanker and cargo aircraft by refueling fighters and simultaneously carry the fighter support personnel and equipment on overseas deployments.

A modified Douglas DC-10, the KC-10A entered service in 1981. Although it retains 88 percent systems commonality with the DC-10, it has additional systems and equipment necessary for its air force mission. These additions include military avionics; aerial refueling boom and hose and drogue; seated aerial refueling operator station; and aerial refueling receptacle and satellite communications.

The KC-10 fleet was modified to add wing-mounted pods to further enhance aerial refueling capabilities. Ongoing modifications include the addition of communications, navigation, and surveillance equipment to meet future civil air traffic control needs, and the incorporation of service bulletins to maintain Federal Aviation Administration certification.

The KC-10A is operated by the 305th Air Mobility Wing, McGuire Air Force Base, New Jersey; and the 60th Air Mobility Wing, Travis AFB. California Air Force Reserve Associate units are assigned to the 349th Air Mobility Wing at Travis, and the 514th Air Mobility Wing at McGuire.

During operations Desert Shield and Desert Storm in 1991, the KC-10 fleet provided in-flight refueling to aircraft from the U.S. armed forces as well as those of other coalition forces. In the early stages of Operation Desert Shield, in-flight refueling was key to the rapid airlift of materiel and forces. In addition to refueling airlift aircraft, the KC-10, along with the smaller KC-135, moved thousands of tons of cargo and thousands of troops in support of the massive Persian Gulf buildup. The KC-10 and the KC-135 conducted about 51,700 separate refueling operations and delivered 125 million gallons (475 million liters) of fuel without missing a single scheduled rendezvous.

The NATO air campaign against Yugoslavia began in March 1999. The campaign dubbed Allied Force culminated after months of preparation. The mobility portion of the operation began in February and was tanker dependent. By early May 1999, some 150 KC-10s and KC-135s deployed to Europe where they refueled bombers, fighters, and support aircraft engaged in the conflict. The KC-10 flew 409 missions throughout the entire Allied Force campaign and continued support operations in Kosovo.

With three General Electric CF6-50C2 turbofan engines producing 52,500 pounds of thrust each, the Douglas KC-10A Extender is nearly twice as powerful as its smaller sibling, the KC-135R. Capable of carrying four times the weight of the KC-135R, the KC-10A can fly 4,400 miles with a total cargo weight, including fuel, of 342,000 pounds, or 11,500 miles without cargo at speeds up to 600 mph. Four units operate the KC-10A, the 305th Air Mobility Wing and Air Force Reserve 514th Air Mobility Wing from McGuire Air Force Base, New Jersey, as well as the 60th Air Mobility Wing and Air Force Reserve 349th Air Mobility Wing at Travis Air Force Base, California.

"Checklist complete, ready to roll." Sitting in his F-16C Viper, Major Derek "Tazz" Routt of the 64th Aggressor Squadron signals his crew chief that all is a go, canopy to close, chocks removed, and ready to begin the long taxi towards the south end of the base. He, along with five other pilots from the 64th, will make up Red Air for this particular mission during Joint Red Flag 2005. JRF2005 has since been the largest joint exercise operation held by the U.S. Air Force with the participation of multiple nations from around the world.

aggressor squadron have now become reality. The 64th and 65th Aggressor Squadrons recently participated in their first Red Flag exercise flying out of Eielson Air Force Base, Alaska.

Space Capable Adversary

In 2001, the 527th Space Aggressor Squadron from Schriever AFB, Colorado, joined Red Air to make life miserable for downed airmen and those looking to support a rescue. Their role was to play the "space capable adversary" and knock out global positioning system (GPS) equipment needed for a successful search-and-rescue mission. So successful was the added operation that Blue Air spent so much time concentrating on the problem at hand that they never even noticed the Red Air aggressors moving in for the kill.

Situated in remote locations on the Nellis AFB range, the 527th and the 14th Test Squadron Reserve aggressors set up their equipment at strategic locations. Their goal was to deny GPS capabilities to downed airmen and rescue helicopters, just as the adversary would deny GPS to a downed American pilot. Forces were sure to use readily available equipment and known adversary capabilities in order to achieve precise realism. Never before had a realistic space threat been played in the traditionally air-oriented training event.

The tasks of the 527th can be divided into five different flights to better enhance the technological experience of the scenario. Those flights include the electronic warfare flight, the imagery exploitation flight, the red attack flight, the space control flight, and the 14th Test Squadron's Air Force Reserve space aggressor flight.

The electronic warfare flight makes airmen see what it is like to operate in an environment where space-based assets have come under attack. Using available or known enemy equipment,

An aggressor F-16 representing the Red Team blasts off from runway 3L at Nellis AFB. In the background is just a small sample of the aircraft present during Red Flag exercises. The original concept of an aggressor squadron was to enable these aircraft to visit other bases and train on-site. Although initially that wasn't to be the case, recent changes to Red Flag brought the 64th Aggressor Squadron to Eielson AFB for the first Red Flag—Alaska held in mid-2006.

they employ the tactics of the enemy to jam GPS and satellite communications networks.

The imagery exploitation flight uses commercial sensors to obtain incredibly detailed and accurate imagery of U.S. forces. Anyone with access to a computer and method of payment can obtain the same images. Aggressors use the images to build their understanding of the battle space and to analyze the force protection vulnerabilities of deploying forces.

The red attack flight coordinates the efforts between the flights and devises a coherent opposition campaign plan. To be credible and realistic, it's not enough to throw adversary capability at exercise participants; the effort must be a coordinated plan with its own battle rhythm, and it should be based on the strategies, doctrine, and tactics of a particular enemy. To do otherwise would reduce the value of the 527th's operations in the exercise.

The space control flight analyzes future countermeasure capabilities and develops new tactics and procedures in case of an attack on U.S. space assets.

Finally, the 14th Test Squadron's Air Force Reserve space aggressor flight supports all aspects of the 527th's mission, providing a variety of operationally experienced people, helping integrate the air and space worlds.

Due to circumstances brought about in the Rumsfeld Space Commission Report and the Air Force Chief of Staff's Aerospace Integration Plan, the ability to respond to space-capable dangers has become a high priority for air force and joint combat exercises. Countries posing a credible threat have proven that space integration is the way of the future in exercises.

Threat Training Facility

The weapons museum at Nellis, or the Threat Training Facility (TTF) as it is officially known, was established in 1976 under what was then the 4513th Tactical Training Group. During the past twenty-three years, the TTF has grown from a collection of a few antiaircraft artillery (AAA) pieces to a large and diverse collection of foreign weapons that include surface-to-air and

Affectionately known as the "Petting Zoo," the threat training lab (TTL) has managed to acquire some of the most elusive enemy weaponry known to exist. Among the collection is this Mil-24 Hind helicopter made fully accessible to pilots and aircrew. Scattered elsewhere throughout the base is an enormous collection of MiG and Sukhoi aircraft dating from the Korean War era to the present day. Also included throughout the compound are models of aircraft that have yet to be acquired, as well as small-arms fire, antiaircraft artillery (AAA), squadron logos, and much more.

A once highly classified facility, the threat training lab (TTL) located at Nellis AFB provides pilots and aircrew with firsthand knowledge of enemy arms. Among the various weaponry accumulated is this mobile surface-to-air missile (SAM) platform. Having the ability to walk around and examine hostile arms enabled warfighters the chance to better understand the machinery they were training to fight against as well as its limitations and weak points.

A museum of adversary weapons, the threat training lab (TTL) contains everything a pilot may encounter over hostile territory. Unfortunately, pilots do occasionally become the victim of hostile fire and face the realization of coming in contact with small arms. Because the goal of the TTL is for pilots to become more familiar with these arms, the lab works hard to acquire a real example of every type of weapon. From enemy attire, badges, and flight suits to adversary aircraft, helicopters, and mobile radar units, the TTL is one of the only places in the world where friendly forces have the chance to touch, try out, and study enemy machinery.

air-to-surface missiles, aircraft, armored vehicles, and numerous types of small arms.

Better known as the "petting zoo," the facility is designed to give pilots an understanding of what they might face in combat and how to defeat it. What started out as a display of about fifteen antiaircraft guns has grown to include weapons systems from all over the world with a total value of more than $100 million. It's the 547th's responsibility to produce the constantly updated Air Force Threat Reference Guide and Countertactics, making the TTF facility invaluable.

The TTF was designed to educate and train aircrews and support personnel from all U.S. services on adversary weapons systems. Up until the latter part of the 1980s, stringent measures were taken to ensure that access was only given to those with a secret clearance. Since that time, the TTF has opened its doors to a wide range of visitors, including active-duty military, VIPs from all branches of the military (U.S. and foreign), civic leaders, and even Russian (former–Soviet Union) dignitaries.

TTF personnel give both classified and unclassified briefings for visiting groups. These individuals stay abreast of the latest information on all the weapon systems in order to provide the most current information to visitors. Some of the most recent acquisitions to the yard include a MiG-29 Fulcrum, MiG-23 Flogger, an Iraqi Mi-24 Hind D helicopter, an actual SA-8 Gecko, and an SA-6 Gainful missile, replacing an existing full-size replica.

In 1993, the TTF was declassified, and now invited civilian guests can tour the facility. More than 31,000 civilians and soldiers tour the TTF every year.

Chapter 4

THE BLUE TEAM

Two F-16CGs from Hill Air Force Base's 388th Fighter Wing, 421st Fighter Squadron, take a subtle left turn showing off the low-altitude navigation and targeting infrared for night (LANTIRN) pod just under the intake. The LANTIRN system allows the F-16 to fly low altitudes at night and under the weather to successfully attack ground targets with precision-guided and unguided weapons. Block 50 F-16CGs are the only night- and all-weather-navigation and precision-attack-capable F-16 models flown and were occasionally dubbed the Night Falcon.

A sixteen-year veteran fighter pilot, Major Mark "T-Bone" Doria has participated in five Red Flag exercises. Even as a recipient of numerous air force commendations including the Defense Meritorious Service Medal, two Meritorious Service Medals and five Air Medals for combat over the Balkans and Iraq, the exercises still prove daunting. "Red Flag is an extremely intense and intimidating environment. Finding your target with so many aircraft and worrying about Red Air makes it very challenging."

Flying with the 421st Fighter Squadron out of Hill Air Force Base, Major Doria makes up just one of the many components of the Blue Force. Typically the Blue Force consists of an air expeditionary force and gathers at Nellis Air Force Base thirty to ninety days prior to deployment. Not only does this give the various elements the proper training and provide combat readiness, but also the ability to get to know those from other units in the same training session.

In order to cover all aspects of potential scenarios, the Blue Force will train in offensive counter air (OCA), defensive counter air (DCA), interdiction (INT), suppression of enemy air defenses (SEAD) and command-and-control (CC). Other aspects such as close air support (CAS), combat search and

A Blue Team F-16 from the 388th Wing, 421st Fighter Squadron, the Black Widows, flies high above the Nevada desert during a training exercise. Typically, Red Flag exercises are worked around the deployment schedules of Air Expeditionary Forces (AEF). Approximately thirty to ninety days before deployment, all applicable units begin training, which helps the various players become more cohesive in their mission objectives. Many claim Red Flag exercises build a better camaraderie between units.

Major Mark "T-Bone" Doria's Experience as "Blue Four"

"I remember being a young pilot, the eerie glow of the setting sun surrounding me as I sat in the cockpit watching the cloud cover begin to increase. In the back of my mind, the worsening weather conditions discussed at the brief wreaked havoc on my concentration. A few moments after starting the engine and completing my checklist, my crew chief gave the all-clear signal. With chocks removed from my F-16 and the engine spooled up, I followed my lead, Captain Glenn 'Greedy' Reedy, now an air force lieutenant colonel, down the very long taxiway.

Starting out as a group of four, the second F-15 climbs to altitude silhouetted against the late evening sky with the Las Vegas skyline below. Following closely behind, the third F-15 is at rotation and a fourth can be seen at the far end of the runway with afterburners lit. Because the Nellis Range Complex is mostly uninhabited, nighttime missions can be extremely difficult with the majority of concentration focused on night-vision technology. Whether it be night-vision goggles (NVGs) or panel-mounted forward-looking infrared (FLIR) devices, night missions teach pilots to trust their instrumentation.

"Glancing over to runway 3R, a pair of Boeing KC-135 Stratotankers took to the sky followed by the brilliant glow of four afterburners from a Rockwell B-1B Lancer bomber. In sharp contrast, a mellow saucer-laden Boeing E-3 Sentry rolled down the runway, all but disappearing if not for its flashing wing-tip strobes. While making the long, slow taxi, I couldn't help but take note of the Las Vegas skyline beginning to replace the glow of the setting sun. Just below my view of the skyline, I see an endless row of F-15s and more F-16s waiting clearance from the Nellis tower at the southwest end of the base.

Kicking off one of the many intense nighttime missions, an RC-135 V/W Rivet Joint electronic surveillance aircraft takes to the skies. Typically, the larger aircraft such as the refuelers, electronic reconnaissance, and bombers head to the Nellis Range Complex prior to the fighters due to increased loiter times. The RC-135 specializes in the gathering of enemy information and relates that knowledge to friendly forces. The aircraft can even speak directly to an aircraft that is in immediate danger of enemy activities.

"A few more aircraft take to the skies ahead of us, a pair of F-117 Nighthawks from Holloman and a four-ship of F-15s from Eglin. Finally we taxied over the numbers and took position wingtip to wingtip looking down the lengthy runway towards the northeast.

"The sky soon became a deep, dark blue, as the cloud cover continued to increase, an illusion from the setting sun. I glanced around the cockpit one last time checking my engine gauges and verifying switches. Off to my left, out of the corner of my eye, the first F-16 began to roll. I looked up to see lead light the afterburner causing the runway to glow from the rising heat of the earlier afternoon's warmer weather.

"With my left hand gripping the throttle and my feet gently applying pressure to the pedals, I nudge the throttle forward, let off the brakes and thundered down runway 3L. Tapping the pedals to keep the Viper straight, I push the throttle one notch further forward and the afterburner kicks in. As other aircraft start up and begin their taxi, all that could be seen to my left is a rushing blur while I gently pull back on the stick. After a minor correction for a subtle crosswind, I head off to the Nellis Range Complex.

"A green glow filled the cockpit from the multitude of panels and knobs, while my ears were filled with chatter from ground control and the intense coordination needed to get eighty-plus aircraft in the air. Watching the heads-up display (HUD) along with the faint blinking lights in the distance, I soon joined up on Reedy and followed him to our predesignated airspace. The thought of so many aircraft in the same area combined with ominous and foreboding weather caused my mind to race. Involuntary stretching of the hands, intense concentration on positioning with lead and numerous quick glances at my HUD were all I could do to keep my mind off of all the variables that could go so wrong.

"Of course I knew I wasn't alone. Every pilot having flown Red Flag for the first time gets that pit-in-the-stomach feeling. Unfortunately, that knowledge didn't help. I hung on to Greedy for dear life as he took us to the target and back."

Today, Major Mark "T-Bone" Doria flies with the 421st Fighter Squadron out of Hill Air Force Base. His sixteen years as a veteran pilot have brought him numerous commendations from combat experiences around the world.

The ghostly silhouette of an F-15 supported by spears of light blasts overhead on its way out to the Nellis Range Complex for a nighttime mission. With full afterburner, the F-15C can produce fifty thousand pounds of thrust utilizing the two Pratt & Whitney F100-PW-100 afterburning turbofan engines. With a nominally loaded weight of forty-eight thousand pounds, the Eagle can accelerate while in a vertical climb and, if straight and level in a clean configuration, can do over two and a half times the speed of sound.

A group of four F-16Cs return from the Nellis Range after completing a sortie. The objective of Red Flag is not only to familiarize pilots with a combat environment, but to introduce Blue Four to these challenging conditions. Seen here flying in left echelon formation, Blue Four is so named for being the least experienced pilot typically flying in the fourth trailing position.

rescue (CSAR), special operations forces (SOF), tactical airlift (TA), and air-to-air refueling (AAR) are integrated into Red Flag as needed to optimize aircrew training. Each two-week session has a predetermined scenario that allows for air-to-air and surface-to-air threat and target-array training.

Call sign "Anchor 21" flies the oval pattern above the Nellis Range Complex. Far below the KC-135R are the groups of aircraft making up both the Red and Blue Teams. This particular KC-135R from the 186th Aerial Refueling Wing from Mississippi has been tasked with serving Blue Air for this mission enabling friendly aircraft the ability to refuel quickly and return to the fight. Located on the opposite side of the range, an additional KC-135R is prepared to assist Red Air.

Most of the concentration is placed on the Blue Four as well as training young mission commanders how to develop complex strike plans. Blue Four is typically the least-experienced pilot and usually flies as the fourth position in an echelon formation. Because Red Flag has the potential of becoming a demoralizing experience, the exercise practices a crawl, walk, and run approach.

Besides concentrating on the newer pilots, Red Flag missions also give mid-level captains the ability to plan a

Allied Aircraft: Nimrod MR2

The Nimrod carries out three main roles:

Antisubmarine Warfare — aerial patrols monitoring the activities of submarines.

Anti-Surface-Unit Warfare—covers a wide range of operations involving surveillance and reconnaissance missions that may culminate in the targeting and attack of enemy vessels.

Search and Rescue—supporting SAR operations by assisting in the detection and location of personnel in emergency situations, providing a communications relay with attending helicopters.

The Nimrod entered service in 1969 as the MR1 version. Based on the civilian Comet airliner, the Nimrod was, and remains, the only jet-powered, long-range, maritime-patrol aircraft in military service. Offering the advantages of speed and height during transit, while still capable of long patrol periods and, in particular, stealth in the antisubmarine mission. (Propeller-driven aircraft make a discrete resonance that can be detected by submerged submarines, whereas the jet noise of the Nimrod is virtually undetectable.)

All Nimrod MR2s are based at RAF Kinloss, equipping No. 120 and 201 Squadrons along with the Operational Conversion Unit, No. 42 (Reserve) Squadron. The majority of Nimrod tasking comes from the Maritime Headquarters (MHQ) at Northwood. Peacetime work includes surface and subsurface surveillance and taking part in maritime exercises around the world; much of the Nimrod's work is in support of naval forces and it is essential to remain current in joint operations. There is always an aircraft on one-hour readiness for search and rescue, primarily for downed military aircrew and military maritime incidents. Nimrods, however, are tasked by the Air Rescue Coordination Center at Kinloss to attend many civil incidents. Such activity may include carrying out searches, assisting search-and-rescue helicopters or acting as on-scene commander at major incidents such as the Piper Alpha oil-rig disaster. The aircraft can operate as low as two hundred feet while over the sea.

A No. 51 Squadron Nimrod R1 launches from runway 27R for a nighttime mission over the Nellis Range Complex. Capable of reaching speeds of nearly six hundred miles per hour, the R1 provides electronic surveillance support to allied fighters with the ability to maintain long loiter times as well as being able to get to the scene quickly. Although the Nimrod was originally designed for antisubmarine and maritime patrol duties, newer R1s were equipped with a highly sophisticated and sensitive suite of systems used for reconnaissance and the gathering of electronic intelligence.

ANG

On the left wingtip pylon of this F-16C stationed at Luke AFB is a special instrument pod known as the Nellis air-combat training system (NACTS). Introduced in 1999, NACTS assists in relaying details valuable to informative debriefings. Capable of tracking up to one hundred separate aircraft, the pod transmits data regarding position, altitude, speed, and weapons use. The individual combat aircrew display system (ICADS), improved fidelity, and GPS technology, as well as increased range, are all welcome improvements.

mission involving around 140 aircraft. As the Red Flag mission commander, they are tasked with coordinating air tankers and weapons assets along with determining priority targets. With assets and coordinating time purposefully lessened, the goal of the Red Flag commander is to destroy the highest number of priority targets with the least amount of losses.

Left: Taking on fuel from a KC-135R from the 141st Aerial Refueling Wing, this F-16 from the 421st Fighter Squadron, the Black Widows, demonstrates a basic weapons load. On stations three and seven reside two two-thousand-pound GBU-31 inactive bombs with stations six and four carrying extended-range fuel tanks. One of the most popular aircraft ever built, the F-16 is utilized by more than twenty-five different countries, including the United States, and was the first fly-by-wire aircraft. Instead of the pilot manually moving the control surfaces with a series of rods and cables, electronic signals are sent from the control stick to the applicable control surfaces.

Pilots are first introduced to the Nellis Range Complex through a local area orientation flight. Once actual exercises begin, everything is taken in steps. On the first day of Red Flag, no bombs are dropped and the aggressor aircraft resemble MiG-29s with less-capable maneuverability and weapons and no electronic attacks. Towards the end of the first week, difficulty is stepped up targeting with inert bombs and maneuvering against more agile aggressors such as the SU-27s. By the second week, tactics resembling the flight characteristics of the highly capable SU-30 MKK and increased surface-to-air activity are in play.

Flying over the Nellis Test and Training Range, pilots can encounter a multitude of targets including trains, industrial complexes, armored vehicles, railways, convoys, airfields, bridges, and high-threat targets such as radar installations, surface-to-air

B-1B Lancer

Carrying the largest payload of both guided and unguided weapons in the air force inventory, the multi-mission B-1 is the backbone of America's long-range bomber force. It can rapidly deliver massive quantities of precision and non-precision weapons against any adversary, anywhere in the world, at any time.

The B-1A was initially developed in the 1970s as a replacement for the B-52. Four prototypes of this long-range, high speed (Mach 2.2) strategic bomber were developed and tested in the mid-1970s, but the program was canceled in 1977 before going into production. Flight testing continued through 1981.

The B-1B is an improved variant initiated by the Reagan administration in 1981. Major changes included the addition of additional structure to increase payload by 74,000 pounds, an improved radar, and reduction of the radar cross-section by an order of magnitude.

The first production B-1 flew in October 1984, and the first B-1B was delivered to Dyess Air Force Base, Texas, in June 1985. Initial operational capability was achieved on October 1, 1986. The final B-1B was delivered May 2, 1988.

The B-1B holds almost fifty world records for speed, payload, range, and time of climb in its class. The most recent records were made official in 2004.

The B-1B was first used in combat in support of operations against Iraq during Operation Desert Fox in December 1998. In 1999, six B-1s were used in Operation Allied Force, delivering more than 20 percent of the total ordnance while flying less than 2 percent of the combat sorties. Eight B-1s were deployed in support of Operation Enduring Freedom (OEF). B-1s dropped nearly 40 percent of the total tonnage during the first six months of OEF. All of this was accomplished while maintaining an impressive 79 percent mission capable rate.

Originally designed as a replacement for the aging Boeing B-52 Stratofortress, the B-1B Lancer is capable of carrying the largest payload of any aircraft in U.S. inventory. Despite the current success of the B-1B, it was almost written off as a failure. In 1981, the Reagan administration sent Rockwell back to the drawing board with the B-1A improving the aircraft's structural integrity and payload while decreasing the radar signature. Although top speed was reduced from Mach 2.2 to 1.2 after airframe enhancements, future plans for the B-1B call for the addition of LINK-16 capability, allowing the B-1B to operate in the integrated battlefield of the future.

One of four different camouflage paint schemes adopted by the 64th Aggressor Squadron adorn this F-16C Viper. Maintaining a relatively low operating cost and providing extreme maneuverability and targeting tactics, the F-16 is the perfect aircraft to assume the aggressor role of the Red Team. To combat similar aircraft from the Blue Team, the 65th Aggressor Squadron composed of F-15Cs has been reestablished and will be flying similar camouflage schemes.

missiles, and antiaircraft armament. Buildings with remote-controlled vehicles patrolling the perimeter surrounded by simulated waterways and roads simply add to the realism.

The ability to confuse an attacker as well as track the precision of the attack is monitored by sophisticated technology. Not only can ground sources detect that a building has been hit, but they can tell precisely which floor and which window the bomb went through. The same goes for the multiple remote-control vehicles roaming the range. Equipped with GPS receivers, airmen in sheet-metal covered buggies task intelligence, reconnaissance, and surveillance personnel with the challenges of finding, tracking, and handing the target off to an attacker.

Each aircraft is equipped with an instrument pod to obtain and transmit data on position, altitude, speed, and weapons use. Known as the Nellis Air Training System (NATS), the pods assist in relaying information valuable to informative debriefings due to a more accurate assessment of engagements. Introduced in 1999, NATS increased the number of high-activity aircraft that can be tracked simultaneously from thirty-six to one hundred. It also increased the number of surface-to-air missile launcher threats from thirty to seventy-five, and doubled the number of simultaneous weapons simulations from fifty to one hundred. The system also features individual combat aircrew display system (ICADS), GPS technology, improved fidelity, and an extended range size.

The Aircraft

High above the Nellis Test and Training Range, Blue Air utilizes almost every air force asset to accomplish their mission. Typically an exercise will consist of two countries embroiled in

Visiting from Tinker Air Force Base in Oklahoma, an E-3 Sentry from the 552nd Air Combat Wing takes off from runway 3R on its way out to the Nellis Range Complex. During Red Flag exercises, most of the larger aircraft such as the heavy bombers, tankers, and those providing electronic support head to the range first in order to best direct the front-line fighters. The E-3 Sentry, with its enormous radome, can scan a 250-square-mile radius, providing up-to-the-minute information on allied and enemy locations.

battle, and it's up to the Blue Team to either invade and free the imaginary nation similar to the scenario involving Kuwait during Desert Storm, or defend their ground against the attacking Red Team.

Just as in any foreign battle, the Red Team has settled into the range and has the upper hand. With the Red Team having SAM, AAA, and radar installations scattered throughout the range, as well as the ability to regenerate fighters when shot down, the Blue Team needs to provide a mobile task force capable of winning the various scenarios presented to them each day.

Information supplied to members of the Blue Team comes from aerial platforms such as the RC-135V/W Rivet Joint, E-8C Joint STARS and E-3 Sentry AWACS aircraft.

The radar and computer subsystems on the Boeing E-3 Sentry airborne warning and control system (AWACS) can gather and present broad and detailed battlefield information. Data is collected as events occur. This includes position and tracking information on enemy aircraft and ships, and location and status of friendly aircraft and naval vessels. The information can be sent to major command-and-control centers in rear areas or aboard ships. In time of crisis, this data can be forwarded to the president and secretary of defense in the United States. It is a jam-resistant system that has performed successful missions while experiencing heavy electronic countermeasures.

The E-8C joint surveillance target attack radar system (Joint STARS) is an airborne battle management, command-

Allied Aircraft: MiG-29

The MiG-29, NATO-codename Fulcrum, is one of Russia's most advanced and capable fighter aircraft. Designed primarily as an air-superiority fighter, it reached operational status with the Soviet air defense forces in 1985. Today, the mission of the MiG-29 is to engage hostile air targets within radar coverage limits, and also to attack ground targets using unguided weapons in visual flight conditions.

The MiG-29 is one the most successful Soviet/Russian military aircraft designs, currently in service in twenty-three different countries, including Bulgaria, Germany, Hungary, India, Iran, Iraq, Malaysia, North Korea, Peru, Poland, Romania, Russia, Syria, Ukraine, and Yugoslavia.

The U.S. Department of Defense and the Ministry of Defense of the Republic of Moldova reached an agreement to implement the Cooperative Threat Reduction accord signed on June 23, 1997, in Moldova. This is a joint effort by both governments to ensure that these dual-use military weapons do not fall into the hands of rogue states. The Pentagon pounced on the Fulcrum C-models after learning Iran had inspected the jets and expressed an interest in adding them to their inventory. Although Iran already flies the less-capable Fulcrum A, it doesn't own any of the more advanced C-models. Of the twenty-one Fulcrums the United States bought, fourteen are the frontline Fulcrum Cs, which contain an active radar jammer in its spine, six older As and one B-model two-seat trainer. This agreement authorized the United States Government to purchase nuclear-capable MiG-29 fighter planes from the government of Moldova.

Occasional participants of Red Flag exercises are German MiG-29 fighters. Flown by German Luftwaffe pilot Lieutenant Colonel Tom Hahn, squadron commander of the 73rd Fighter Wing or "Steinhoff," he flies in formation with Major Greg Thomas in an F-15C from the 28th Test Squadron. Hosting MiG-29s enable U.S. pilots to fully understand flight characteristics and tactics of equal aircraft operated by hostile nations. *USAF*

The massive Boeing B-52H Stratofortress is one of the oldest aircraft still in service with the U.S. Air Force. Capable of flying at high subsonic speeds at altitudes of up to fifty thousand feet, the Stratofortress can deliver both conventional and nuclear weapons with incredible precision. Combined with the ability for aerial refueling, the B-52's endurance is only limited by its crew. Although the B-52 can perform its duties with minimal input from local forces, the large bomber is still incorporated into Red Flag exercises assisting Blue Air in achieving various tasks throughout the two-week course. *USAF*

Flying at extremely high altitudes, the B-2 stealth bomber has been at the forefront of setting great expectations. Part of the 509th Bombing Wing at Whiteman AFB, Missouri, these bombers are tasked with flying extreme distances. During bombing runs over Afghanistan, B-2s were flying halfway across the world and back, incurring thirty-six-hour-long missions between a crew of two. High over the Nellis Range Complex, two B-2A Spirit stealth bombers await clearance to drop precision-guided weaponry in support of the front-line fighters miles below.

and-control, intelligence, surveillance, and reconnaissance platform. Its primary mission is to provide theater ground and air commanders with ground surveillance to support attack operations and targeting that contributes to the delay, disruption, and destruction of enemy forces. The most prominent external feature is the 27-foot (8 meters) long, canoe-shaped radome under the forward fuselage that houses the 24-foot (7.3 meters) long, side-looking phased-array antenna.

Based on an extensively modified Boeing C-135, the RC-135V/W Rivet Joint's modifications are primarily related to its onboard sensor suite, which allows the mission crew to detect, identify, and geolocate signals throughout the electromagnetic spectrum. The mission crew can then forward gathered information in a variety of formats to a wide range of consumers via Rivet Joint's extensive communications suite.

These aircraft are critical not only to the low-altitude fighters, but also to the high-flying bombers in accurately striking their designated targets. Gone are the days of carpet bombing and citywide destruction. Today's technology has brought with it the expectation of precision attacks with minimal loss to civilian lives.

The oldest aircraft in the air force's inventory is still the most potent. The Boeing B-52 Stratofortress is a heavy long-range bomber that can perform a variety of missions. The bomber is capable of flying at high subsonic speeds at altitudes of up to fifty thousand feet and can carry nuclear or precision-guided conventional ordnance with worldwide precision navigation capability.

The aircraft's flexibility was evident in Operation Desert Storm and again during Operation Allied Force. B-52s struck wide-area troop concentrations, fixed installations and bunkers, and decimated the morale of Iraq's Republican Guard. The Gulf War involved the longest strike mission in the history of aerial warfare when B-52s took off from Barksdale Air Force Base, Louisiana, launched conventional air-launched cruise missiles, and returned to Barksdale—a thirty-five-hour nonstop combat mission.

Almost a complete opposite to the B-52, the Rockwell B-1B's blended wing/body configuration, variable-geometry

B-2 Spirit

The B-2 Spirit is a multi-role bomber capable of delivering both conventional and nuclear munitions. A dramatic leap forward in technology, the bomber represents a major milestone in the U.S. bomber modernization program. The B-2 brings massive firepower to bear, in a short time, anywhere on the globe through previously impenetrable defenses.

The first B-2 was publicly displayed on November, 22, 1988, and its first flight was July 17, 1989. The B-2 Combined Test Force, Air Force Flight Test Center, Edwards Air Force Base, California, is responsible for flight testing the engineering, manufacturing, and developing B-2 aircraft.

Whiteman AFB, Missouri, is the only operational base for the B-2. The first aircraft, *Spirit of Missouri*, was delivered December 17, 1993.

The combat effectiveness of the B-2 was proved in Operation Allied Force, where it was responsible for destroying 33 percent of all Serbian targets in the first eight weeks, by flying nonstop to Kosovo from its home base in Missouri and back. In support of Operation Enduring Freedom, the B-2 flew one of its longest missions to date from Whiteman to Afghanistan and back. The B-2 completed its first-ever combat deployment in support of Operation Iraqi Freedom, flying twenty-two sorties from a forward operating location as well as twenty-seven sorties from Whiteman AFB and releasing more than 1.5 million pounds of munitions.

The prime contractor, responsible for overall system design and integration, is Northrop Grumman Integrated Systems Sector. Boeing Military Airplanes Company, Hughes Radar Systems Group, General Electric Aircraft Engine Group, and Vought Aircraft Industries, Inc., are key members of the aircraft contractor team.

Just as the United States Navy has done for over two hundred years, the air force began commissioning certain aircraft by giving them individual names. Here a B-2A Spirit stealth bomber, *Spirit of Kitty Hawk*, lifts off from runway 3R at Nellis Air Force Base. For this particular exercise held in early 2006, four B-2A bombers from the 509th Bombing Wing, normally stationed at Whiteman AFB, attended providing ground support to advancing warfighters.

At an average cost of $2 billion by 1998 standards, the Northrop Grumman B-2A stealth bomber is costlier than a U.S. Navy aircraft carrier and costlier than just under double its weight in gold. Only twenty-one B-2As were built, all stationed at Whiteman Air Force Base in Missouri under the command of the 509th Bombing Wing. Despite the limited production of the YB-49 Flying Wing in the late 1940s, the B-2A is the first flying-wing aircraft to enter service.

wings, and turbofan afterburning engines combine to increase range, maneuverability, and speed, while enhancing survivability. The B-1B's speed and superior handling characteristics allow it to seamlessly integrate in mixed-force packages. These capabilities, when combined with its substantial payload, excellent radar-targeting system, long loiter time, and survivability, make the B-1B a key element of any joint/composite strike force. The B-1 weapon system is capable of creating a multitude of far-reaching effects across the battlefield.

The B-1A was initially developed in the 1970s as a replacement for the B-52. Four prototypes of this long-range, high-speed (Mach 2.2) strategic bomber were developed and tested in the mid-1970s, but the program was canceled in 1977 before going into production. Flight testing continued through 1981.

Although the B-1 series almost didn't make it, the potential of the aircraft was eventually noticed. The production B-1B was first used in combat in operations against Iraq during Operation Desert Fox in December 1998. In 1999, six B-1s were used in Operation Allied Force, delivering more than 20 percent of the total ordnance while flying less than 2 percent of the combat sorties. Eight B-1s were deployed in support of Operation Enduring Freedom (OEF). B-1s dropped nearly 40 percent of the

A unique view of four F-16C Fighting Falcons surrounding a KC-135R tanker, call sign Anchor 21 from the 186th Aerial Refueling Wing from Mississippi. Air force aircraft use the probe system for aerial refueling in place of the drogue system used by the navy. A long probe is extended aft the KC-135 equipped with small wings that are controlled by the boomer. The boomer will swing the probe into position using these wings while the F-16 pilot maneuvers his aircraft as close to the boom as possible. Each aircraft will assume a position alongside the tanker taking turns receiving fuel.

total tonnage during the first six months of OEF. This included nearly 3,900 joint direct-attack-munition (JDAM) bombs, or 67 percent of the total. All of this was accomplished while maintaining an impressive 79 percent mission capable rate.

Along with the B-52 and B-1B, the Northrop Grumman B-2 Spirit, often called the stealth bomber, provides the penetrating flexibility and effectiveness inherent in manned bombers. Its low-observable, or stealth, characteristics give it the unique ability to penetrate an enemy's most sophisticated defenses and threaten its most valued, and heavily defended, targets. Its capability to penetrate air defenses and threaten effective retaliation provides a strong, effective deterrent and combat force well into the twenty-first century.

The combat effectiveness of the B-2 was proved in Operation Allied Force, where it was responsible for destroying 33 percent of all Serbian targets in the first eight weeks, by flying nonstop to Kosovo from its home base in Missouri and back. In support of Operation Enduring Freedom, the B-2 flew one of its longest missions to date from Whiteman, Missouri, to Afghanistan and back. The B-2 completed its first-ever combat deployment in support of Operation Iraqi Freedom, flying twenty-two sorties from a forward-operating location as well as twenty-seven sorties from Whiteman AFB, and releasing more than 1.5 million pounds of munitions. The B-2's proven combat performance led to declaration of full operational capability in December 2003.

Left: A mandatory requirement for an efficient mission is the ability to quickly refuel while over the range. During combat tactics, aircraft such as this F-15C from the 58th Fighter Squadron based at Eglin AFB, Florida, burn fuel extremely fast and must find the tanker, fuel, and return to the battlefield as quickly as possible. Typically, a group of two to four aircraft will approach the tanker and form up as they all maintain an oval pattern over a designated area. As soon as one aircraft has been fueled, the next slides in and connects. Not until all aircraft in the group are fueled, do they depart the tanker and head back to the battlefield

"Chick Flights"

After an eight-hour drive from my Monterey home, I'm at the front gate of Nellis Air Force Base, watching the sunrise above the heroic Thunderbirds statue. I start the phone calls necessary to contact my designated aircrew, but a lot of people aren't on base yet. Fortunately, a chief from the 141st Air Refueling Wing based at Fairchild, Washington, was already hard at work.

"Mr. Rininger, am I correct to assume you were hoping to fly today?" asked the chief, slightly puzzled.

"Yes, sir. Those were my plans."

"Well, engines are already turning. I'm not sure if I can get you out here in time." There was a long pause. "Tell you what, are you ready to go? Like now!?"

"Yes, sir. Just need someone to pick me up."

"I'll be right there."

Like in a Tom Clancy novel, I'm whisked onto the base and out to the flightline, where the crew chief and aircraft maintenance personnel work hard at preflighting one of the three KC-135s in the morning sortie. Clumsily exiting the Suburban with my camera gear, I tried to multitask, putting in earplugs while paying attention to hand signals as the roar on the flightline got louder and louder.

From left to right: Lieutenant Colonel Patricia Morales, Captain Molly Marshall, and Master Sergeant Sheri Shaw.

A KC-135 from the 141st Air National Guard Aerial Refueling Wing based at Fairchild, Washington, takes to the sky to begin the nighttime Red Flag exercise.

Below and aft the cockpit, a ladder extends downward allowing cabin access. I hand my camera gear to a maintainer already in the aircraft and climb up to the flight deck. With camera equipment stowed, I hardly noticed Master Sergeant Sheri Shaw preparing the jump seat for me, a rickety folding chair in the center of the cabin with the best view in every direction.

An F-16CG approaches the boom to take on fuel. Blue Team F-16s typically arrive in groups of four; two aircraft on each side of the KC-135R rotate through to make the transfer as quick as possible. This F-16 is from the 388th Fighter Wing, 421st Fighter Squadron, based at Hill Air Force Base in Utah.

I hear the hatch close behind me and, at the same time, am handed a headset to listen in on the pre-flight checklist. Lieutenant Colonel Patricia Morales, the pilot, sits in the left seat, with Captain Molly Marshall acting as copilot and navigator. Now I've never been the sharpest knife in the drawer, but suddenly it dawned on me that I was the only guy on this plane. Although I've been on plenty of tanker flights, this was my first all-female flight crew. Not knowing if this was a new thing or typical for the Air Guard, I kept my mouth shut and enjoyed the takeoff.

After about twenty minutes in the air, the flight crew began talking about how cool this all-female flight was. After a surprised inquiry, I found this to be the first all-female TDY flight for the 141st ARW. The crew named it the "chick flight" and went on to call it the unit's first "unmanned aerial refueling mission." While maintaining the utmost professionalism, the flight's atmosphere was lighter than any I had been on before, a pleasant relief from the usual rigid practices.

Throughout the Red Flag exercise, the chick flights continued and became the talk of the base. More than just your average flight crew, Morales, Marshall, and Shaw uphold an incredible lifestyle not only serving their country, but taking on the challenges of being mothers and wives.

Once I got to know them better, I understood why the flight routine was so relaxed. A wife of seventeen years and a mother of five, Lieutenant Colonel Morales began flying at the age of seventeen and acquired her commercial instrument rating while still in college. Joining the Air Force after college, she first flew the KC-135A in 1986. Ten years later, she separated from active duty and flew C-141s with the 313th Air Lift Squadron at McChord AFB for three years, ultimately moving back to the KC-135R at Fairchild AFB near Spokane, Washington.

Captain Molly Marshall, flying right seat on our first flight, is also a qualified KC-135R pilot and took the left seat on our second hop. "I'm a Guard Bum," Marshall says, serving with the Air National Guard for nineteen years. Her experience with KC-135s began in 1994 as she trained to become a boom operator. In 1998 she was selected for pilot training, and in late 1999 she began flying the KC-135R. Married and raising a two-year-old daughter, she found out another child was on the way only a week after participating in the first Red Flag exercise of 2006.

With an estimated 4,300 hours of flight time and nearly eighteen years in the Air National Guard, Master Sergeant Sheri Shaw is a boom operator. "I remember this time . . . an F-14 Tomcat approached and plugged in. The pilot and radar intercept officer were really chatty and having a good time. They told me they were aspiring musicians and wanted to see what I thought of this new song they were working on. 'Sure, go for it,' I said."

To the tune of Sade's 1980s hit, "Smooth Operator," the Tomcat crew sang, " 'Cause she's a boom operatorrrrr, a boom . . . operatorrrrr." A newlywed and hopeful mom, Shaw plans on serving another twenty years—a goal of thirty-eight years total before retiring. No doubt she'll have many more stories by then.

Marshall says they're more than just a flight crew. "We play volleyball and get our families together quite a bit. When you work with people for so long, you become a family and it seemed natural to forge our relationships outside of work. We seem to be there for all the big events in life, marriages, babies, promotions . . . the bad stuff too. We stick by one another through thick and thin!"

Lieutenant Colonel Patricia Morales reaches for the communications panel prior to landing at Nellis AFB and concluding the mission. Without reverse-thrust capabilities of newer aircraft, landing the heavy KC-135 can be a challenge.

Captain Molly Marshall taps the throttles to stay on the oval flightpath above the Nellis Range Complex. During the exercise, KC-135 tankers fly predetermined routes awaiting fighters from the Red Team and Blue Team in need of fuel.

Master Sergeant Sheri Shaw rests her chin while cautiously guiding the refueling probe to an F-16. Located in the rear belly of the KC-135, the boomer lies flat while controlling two wings half way up the refueling probe, "flying" the boom into the receptacle.

Silhouetted against the Las Vegas skyline in a surreal irony of events, the normally invisible black F-117A Nighthawk can be seen resting atop its very own heat plume just beneath the Rio Hotel in the background. Able to carry two two-thousand-pound bombs in an enclosed bay, the F-117 is able to remain undetected for almost the entire duration of the mission. Currently there are fifty-four F-117s in service stationed at Holloman Air Force Base in New Mexico.

Supporting the heavy hitters as well as the surveillance aircraft are the KC-135 and KC-10 tankers. Among the oldest aircraft still flying with the air force, the Boeing KC-135 Stratotanker's principal mission is air refueling. This unique asset greatly enhances the USAF's capability to accomplish its primary missions of global reach and global power. It also provides aerial refueling support to U.S. Air Force, Navy, and Marine Corps aircraft as well as aircraft of allied nations. The first aircraft flew in August 1956, and the initial production Stratotanker was delivered to Castle Air Force Base, California, in June 1957. The last KC-135 was delivered to the air force in 1965.

Nearly all internal fuel can be pumped through the tanker's flying boom, the KC-135's primary fuel-transfer method. A special shuttlecock-shaped drogue attached to and trailing behind the flying boom may be used to refuel aircraft fitted with probes. An operator stationed in the rear of the plane controls the boom. A cargo deck above the refueling system can hold a mixed load of passengers and cargo. Depending on fuel-storage configuration, the KC-135 can carry up to 83,000 pounds of cargo.

The larger and newer McDonnell Douglas KC-10 Extender is an air mobility command advanced tanker and cargo aircraft designed to provide increased global mobility for U.S. armed forces. Although the KC-10's primary mission is aerial refueling, it can combine the tasks of a tanker and cargo aircraft by refueling fighters and simultaneously carry the fighter support personnel and equipment on overseas deployments.

Right: The McDonnell Douglas KC-10 Extender can provide nearly twice the amount of fuel as the aging KC-135R tanker aircraft. Entering into service in 1981, the KC-10 has proven itself invaluable assisting aircraft from the air force and the navy as well as many allied forces during Operation Enduring Freedom. Although fuel is transferable by either probe or by an optionally attached drogue, fifteen aircraft are modified with two wing-mounted air refueling pods, which allow for simultaneous operations with probe-equipped aircraft.

U.S. AIR FORCE
305TH AMW
514TH AMW

A rear view of an F-117A Nighthawk shortly after takeoff shows the heat-dispersing exhaust designed to reduce the heat signature sought out by heat seeking missiles and infrared-vision scopes. One of the military's best-kept secrets, the F-117A Nighthawk has seen action all over the world, and many speculate it will soon be put to rest. According to proposed defense budgets, the F-117 is scheduled to be retired by the end of 2007 in favor of the F-22 Raptor. Because Red Flag operations have not yet involved the Raptor into the curriculum, stealth aircraft from the 49th Fighter Wing, 8th Fighter Squadron, are still heavily relied upon.

In addition to the three standard main-wing fuel tanks, the KC-10 has three large fuel tanks under the cargo floor, one under the forward lower cargo compartment, one in the center wing area and one under the rear compartment. Combined capacity of the six tanks is more than 356,000 pounds of fuel, almost twice as much as the KC-135 Stratotanker.

The smaller front-line fighters who depend on these support aircraft round out the aerial complement of participating aircraft. Working together, the F-117, F-15, F-16, and A-10 utilize their independent roles to penetrate the enemy or protect their assigned assets based on their given tasks.

The Lockheed Martin F-117A Nighthawk was the world's first operational aircraft designed to exploit low-observable stealth technology. This precision-strike aircraft penetrates high-threat airspace and uses laser-guided weapons against critical targets. The F-117A can employ a variety of weapons and is equipped with sophisticated navigation and attack systems integrated into a digital avionics suite that increases mission effectiveness and reduces pilot workload. Detailed planning for missions into highly defended target areas is accomplished by an automated mission planning system developed specifically to maximize the unique capabilities of the F-117A.

F-22 Raptor

The F-22A Raptor is the air force's newest fighter aircraft. Its combination of stealth, supercruise, maneuverability, and integrated avionics, coupled with improved supportability, represents an exponential leap in warfighting capabilities. The Raptor performs both air-to-air and air-to-ground missions, allowing full realization of operational concepts vital to today's air force.

The F-22A, a critical component of the global strike task force, is designed to project air dominance, rapidly and at great distances, and defeat enemy threats. The F-22A cannot be matched by any known or projected fighter aircraft.

The advanced tactical fighter entered the demonstration and validation phase in 1986. The prototype aircraft (YF-22 and YF-23) both completed their first flights in late 1990. Ultimately the YF-22 was selected as best of the two and the engineering and manufacturing development effort began in 1991 with development contracts to Lockheed/Boeing (airframe) and Pratt & Whitney (engines). Engineering manufacturing design (EMD) included extensive subsystem and system testing as well as flight testing with nine aircraft at Edwards Air Force Base, California. The first EMD flight was in 1997 and at the completion of its flight test life, this aircraft was used for live-fire testing.

The program received approval to enter low rate initial production in 2001. Initial operational and test evaluation by the Air Force Test and Evaluation Center was successfully completed in 2004. Based on maturity of design and other factors, the program received approval for full-rate production in 2005. Air Education and Training Command and Air Combat Command are the primary air force organizations flying the F-22A. The aircraft designation was the F/A-22 for a short time before being renamed F-22A in December 2005.

At a cost of approximately $338 million per aircraft, the F-22 is by far the most expensive fighter in the U.S. inventory. Its sleek lines and unique exterior coating enable the aircraft to have an extremely low radar cross-section. Special heat-absorbent tiles in the exhaust assembly combined with supercruise technology decrease the heat signature, thereby reducing the threat of being hit by infrared or heat-seeking missiles. Four internal weapons bays also keep the aircraft stealthy.

This F-15E is from the 90th Fighter Squadron operating under the 3rd Wing stationed at Elmendorf AFB in Alaska. Although seen here taking off from runway 3L at Nellis AFB, fighter pilots from Alaska are beginning to feel right at home with the newly incorporated Red Flag—Alaska replacing the already successful Cope Thunder exercises. Many aspects of Cope Thunder will remain unchanged despite the new name, but the name change will provide for a more integrated training environment by taking what was learned from Joint Red Flag held in 2005.

The McDonnell Douglas F-15 Eagle is an all-weather, extremely maneuverable tactical fighter designed to permit the air force to gain and maintain air supremacy over the battlefield. The Eagle's air superiority is achieved through a mixture of unprecedented maneuverability and acceleration, range, weapons, and avionics. It can penetrate enemy defense and outperform and outfight any current enemy aircraft. The F-15 has electronic systems and weaponry to detect, acquire, track, and attack enemy aircraft while operating in friendly or enemy-controlled airspace. The weapons and flight control systems are designed so one person can safely and effectively perform air-to-air combat.

Left: Although this F-15C certainly doesn't look its age, the Eagle is slowly being replaced by the stealthy and incredibly maneuverable F-22 Raptor. For over thirty years the F-15C and its newer all-weather variant, the F-15E Strike Eagle, have been the premier air force fighters, smashing all altitude and speed records. Approximately 1,150 F-15s are still in service worldwide with the U.S. Air Force and Air National Guard units as well as the air forces of Israel, Japan, and Saudi Arabia.

The first F-15A flight was made in July 1972, and the first flight of the two-seat F-15B (formerly TF-15A) trainer was made in July 1973. The first Eagle, an F-15B, was delivered to the air force in November 1974. In January 1976, the first Eagle destined for a combat squadron was delivered.

Similar in appearance to the C-model, the F-15E Strike Eagle is a dual-role fighter designed to perform air-to-air and air-to-ground missions. An array of avionics and electronics

A head-on view of an F-16C with subtle vortices flowing over the wings from a gentle banking turn. Towards the end of a successful mission, this F-16C from Hill Air Force Base begins its approach to the tanker to fill up before heading back to Nellis AFB. A closer look at the aircraft shows that it's carrying a NACTS pod on store eight, an LAU 5003 rocket launcher on store seven, extended-range fuel tanks on stores six and four, and an AIM-9 Sidewinder on store two. Not visible is the LANTIRN pod beneath the intake on store five-right.

systems gives the F-15E the capability to fight at low altitude, day or night, and in all weather conditions.

The aircraft uses two crewmembers, a pilot and a weapon systems officer. Previous models of the F-15 are assigned air-to-air roles; the "E" model is a dual-role fighter. It has the capability to fight its way to a target over long ranges, destroy enemy ground positions, and fight its way out. The low-altitude navigation and targeting infrared for night (LANTIRN) system allows the aircraft to fly at low altitudes, at night, and in any weather condition, to attack ground targets with a variety of precision-guided and unguided weapons. The LANTIRN system gives the F-15E unequaled accuracy in weapons delivery day or night and in poor weather, and consists of two pods attached to the exterior of the aircraft.

A compact multi-role fighter aircraft, the General Dynamics/Lockheed Martin F-16 is highly maneuverable and has proven itself in air-to-air combat and air-to-surface attack. It

Closing in on a KC-135R tanker is this fully loaded F-16CJ from the 52nd Fighter Wing, 23rd Fighter Squadron, stationed out of Spangdahlem Air Base in Germany. Demonstrating weapons and technology only a CJ version F-16 can carry, this Falcon shows off a Link 16 SNIPER XR advanced targeting pod (ATP) as seen under the intake on store five-right, as well as two AGM-88 HARM II antiradiation missiles on stores three and seven. On stores one and nine, this F-16 carries two AIM-120 advanced medium-range air-to-air missiles (AMRAAM), while on store two rests an AIM-9 Sidewinder, and finally on store eight, the Nellis air combat training system (NACTS) pod. Meanwhile, the pilot is equipped with the new joint helmet-mounted cueing system (JHMCS), which allows the pilot the ability to point his head at the target whereby the weapons will be directed to follow.

provides a relatively low-cost, high-performance weapon system for the United States and allied nations.

In an air combat role, the F-16's maneuverability and combat radius (distance it can fly to enter air combat, stay, fight, and return) exceed that of all potential-threat fighter aircraft. It can locate targets in all weather conditions and detect low-flying aircraft in radar ground clutter. In an air-to-surface role, the F-16 can fly more than 500 miles (860 kilometers), deliver its weapons with superior accuracy, defend itself against enemy aircraft, and return to its starting point. All-weather capability allows it to accurately deliver ordnance during non-visual bombing conditions.

Since September 11, 2001, the F-16 has been a major component of the combat forces committed to the Global War on Terror, flying thousands of sorties in support of operations Noble Eagle (Homeland Defense), Enduring Freedom in Afghanistan, and Iraqi Freedom.

Flying at extreme low altitudes, the Fairchild Republic A/OA-10 Thunderbolt II is the first air force aircraft specially designed for close air support of ground forces. They are simple, effective, and survivable twin-engine jet aircraft that can be used against all ground targets, including tanks and other armored vehicles.

Flying high over the Nellis Range Complex is a Fairchild A/OA-10 Thunderbolt II from the 52nd Fighter Wing, 81st Fighter Squadron, from Spangdahlem Air Base in Germany. Located within the range are specific areas for A-10 pilots to target outside of Red Flag missions. During Red Flag, A-10s are responsible for clearing ground threats such as tanks, radar facilities, and antiaircraft arms. Between its massive 30mm GAU-8/A Gatling gun and the ability to fire Maverick antitank missiles, not much can withstand the punch of an A-10.

Capable of surviving a direct hit from armor-piercing and high-explosive projectiles up to 23mm, the A/OA-10 Thunderbolt II employs self-sealing fuel cells that are protected by internal and external foam. Manual systems back up their redundant hydraulic flight-control systems. This permits pilots to fly and land when hydraulic power is lost. The Thunderbolt II can be serviced and operated from bases with limited facilities near battle areas. Many of the aircraft's parts are interchangeable left and right, including the engines, main landing gear, and vertical stabilizers.

Literally built around its gun, the Thunderbolt II's 30mm GAU-8/A Gatling gun can fire 3,900 rounds a minute and can defeat an array of ground targets, including tanks. The Thunderbolt II also includes an inertial navigation system, electronic countermeasures, target-penetration aids, self-protection systems, and AGM-65 Maverick and AIM-9 Sidewinder missiles.

The first production A-10A was delivered to Davis-Monthan Air Force Base, Arizona, in October 1975. It was designed specifically for close air support missions and had the ability to combine large military loads, long loiter, and wide combat radius, which proved to be vital assets to the United States and its allies during Operation Desert Storm and Operation Noble Anvil in Kosovo.

Even when the A/OA-10A Thunderbolt II is surrounded by a breathtaking sunset, the aircraft still conveys the looks only a mother could love. Built for survival, the wheels can roll in their pods, which lets the plane perform belly landings without significant damage to the aircraft. Dual engines are mounted away from the Warthog's fuselage; if one is destroyed, the other can propel the craft to safety. Dual vertical stabilizers shield the hot exhaust from Russian-designed heat-seeking missiles, and the long low-set wings are designed to allow flight, even if half a wing is completely blown off.

Though not yet introduced to Red Flag exercises, the Lockheed Martin F-22A Raptor is the air force's newest fighter aircraft. Its combination of stealth, supercruise, maneuverability, and integrated avionics, coupled with improved supportability, represents an exponential leap in fighting capabilities. The Raptor performs both air-to-air and air-to-ground missions allowing full realization of operational concepts vital to the twenty-first-century air force.

The F-22A's characteristics provide a synergistic effect ensuring F-22A lethality against all advanced air threats. The combination of stealth, integrated avionics, and supercruise drastically shrinks surface-to-air missile engagement envelopes and minimizes enemy capabilities to track and engage the F-22A. The combination of reduced observability and supercruise accentuates the advantage of surprise in a tactical environment.

The F-22A will have better reliability and maintainability than any fighter aircraft in history. An F-22A squadron will require less than half as much airlift as an F-15 squadron to deploy. Increased F-22A reliability and maintainability pays off in less manpower to repair and maintain.

Assisting pilots downed during forward combat is the invaluable HH-60G Pave Hawk, a modified version of the army Black Hawk helicopter, featuring an upgraded communications and navigation suite. Its primary mission is conducting day or night operations into hostile environments to recover downed aircrews or other isolated personnel during war. Because of its

A Lockheed Martin Aeronautical Systems F-22A Raptor from the 1st Fighter Wing, 27th Fighter Squadron in full review. Being the most advanced fighter in the U.S. inventory, the Raptor wasn't exactly welcomed into Red Flag exercises. Scheduled to participate fully in the August 2006 Red Flag, the Raptors have been, for the most part, observers. Much will be learned and no doubt new tactics will be devised when the Raptor makes its combat appearance.

A welcome visitor to Red Flag, these RF-111C Aardvarks flew to the Nevada desert from Australia as part of the No. 1 Squadron based at Amberley. Australia is the last country to fly the F-111 airframe and hopes to continue flying the Aardvark through 2010 and quite possibly 2012 pending development of the multiservice F-35 joint strike fighter. Currently, there are thirty-five F-111 aircraft divided between No. 1 and 6 Squadrons, both out of RAAF Amberley.

Launching from runway 3L, is a GR4 Tornado from the No. IX Squadron stationed at Royal Air Force Base Marham, England, on its way out to the Nellis Range Complex. Located on the outboard right wing pylon is a Phillips BOZ-107 chaff/flare dispenser pod that enables the aircraft to remain in a hostile area for an extended amount of time with an additional degree of protection. To defeat surveillance radars, chaff can be employed to provide a safe corridor for following aircraft, and to create false targets that confuse ground defense systems. Used in conjunction with evasive maneuvers, chaff is also used as a self-protection device against surface-based and missile-mounted tracking radars.

versatility, the HH-60G is also tasked to perform military operations other than war. These tasks include civil search and rescue, emergency aeromedical evacuation, disaster relief, international aid, counterdrug activities, and NASA space shuttle support.

One of the exercises performed during Red Flag is the rescue of a downed pilot behind enemy lines. The Sikorsky HH-60G is crucial in making this portion of the exercise possible. Initially, the downed pilot is either dropped off via the Pave Hawk or parachutes from a Lockheed C-130 cargo plane. It's the pilot's responsibility to evade air and ground forces until the Blue Team can penetrate enemy lines and coordinate a pickup. Once the pilot has been located and air support has been provided, the HH-60G Pave Hawk will initiate rescue procedures and evacuate the pilot. Since every aspect of Red Flag is as realistic as possible, the pilot is treated like an enemy imposter until identification can be verified.

Dispersed among U.S.-inventory aircraft and also belonging to the Blue Team, allied nations from around the world participate in the two-week-long Red Flag exercises. Utilizing the same airframes with various modifications, countries such as Spain train their pilots in F-16s and Australia trains their pilots in F/A-18 Hornets. Australia recently sent over a small group of soon-to-be-retired General Dynamics F-111 Aardvarks, the last F-111s flying anywhere in the world.

These are just some of the resources used by the U.S. Air Force and by friendly forces throughout the world to better accomplish their assigned missions. Red Flag assembles the personnel and equipment needed to better train everyone to perform as a team and provide the most realistic scenarios imaginable.

Chapter 5

THE FUTURE OF RED FLAG

Two groups of four F-15C Eagles prepare to depart runway 3L with the ever-growing Las Vegas skyline in the background. A typical mission during Red Flag can consist of more than seventy aircraft, including fighters, bombers, aerial refuelers, electronic countermeasures, aggressors, airborne command-and-control, and others. Though the majority of aircraft occupying the range are fighters, it takes all of these various support aircraft to ensure a successful and realistic mission.

Red Flag has undoubtedly proven itself to be one of the most successful training exercises ever conducted. Although its success didn't happen overnight, given time, the air force has managed to establish a training process adopted by forces throughout the world. Red Flag's greatest ability has been recognizing and utilizing emerging standards of training. Regardless of its success, today's Red Flag is only the beginning.

In mid-March 2005, the Department of Defense chose to once again raise the bar for combat tactical training. Joint Red Flag (JRF) was an all-new exercise featuring the integration of live, virtual, and constructive elements into a seamless joint campaign to challenge warfighters and test command-and-control procedures, processes, and architecture.

Held from March 14 to April 2, 2005, at Nellis Air Force Base, Joint Red Flag incorporated just about every weapon system in the military arsenal: aircraft, ground vehicles, staff, and communications and computer systems. Costing nearly $21 million, and utilizing 161 aircraft and over ten thousand military personnel from forty-four different sites across the country, JRF was the largest Department of Defense exercise in 2005.

Joint Red Flag had the same primary goal as Red Flag, but on a greater level—involving all forces of the military. Combining air and ground forces, including SAM sites, infantry, aircraft that exist only in computers, troop insertion, foreign involvement, and more, JRF 2005 took the best of every training platform and assembled it all in one place. Never before has the integration of computer communications, simulated flight operations, and ground-control movement been so successfully choreographed in one single exercise.

"During the exercise, we have [about] 350 to 400 live joint and coalition sorties each day," said Lieutenant Colonel James Murray, Twelfth Air Force project officer. "That isn't really out of the norm; however, when you combine that with the more than 600 to 700 constructive sorties and 850 virtual sorties we're flying, it's very busy."

"Joint Red Flag is one of the largest and most complex modeling and simulation architectures created to date," said Lieutenant Colonel Mark D. Horn, commander of the 505th Exercise Control Squadron at Hurlburt Field, Florida. "The exercise has successfully linked thirty-one distributed sites, incorporating thirty-four constructive simulations, and over eighteen virtual simulations and weapons tactics trainers. This is a great undertaking and accomplishment that we are all proud of."

The Joint Red Flag scenario overlaid on the southwestern United States a fictional nation, "Heartland," stretching from

Aircraft commander Captain Dean Peterson and copilot Captain Donna Mae Chun from the 22nd Aerial Refueling Wing based at McConnell AFB, Kansas, bank nearly 45 degrees to line up on approach for runway 27L. The crew from McConnell AFB was thrust into the support role for the largest air force exercise for the 2005 fiscal year. Joint Red Flag 2005 was an enormous step forward in the advancement of joint operations and realistic combat training.

United States Air Force Chief of Staff General John Jumper sits in the left seat of a B-25 bomber during an F-106 dedication ceremony. Qualified in almost every air force inventory aircraft including the F-22 Raptor, General Jumper often visits Red Flag exercises. During Joint Red Flag 2005, General Jumper designated the combined air operations center (CAOC) an actual weapons system comparable to that of a bomber or fighter aircraft.

Las Vegas in the west to El Paso, Texas, in the east. Nearby in the Mojave Desert was the tiny, friendly nation "Enclave." Heartland was flanked by two hostile states, "Eureka" in the California area and "El Dorado," which encompassed eastern New Mexico and most of Texas. For purposes of this joint exercise, which included navy participation, much of Mexico was treated as water. The mission was to defend friendly territory and destroy the enemy's long-term ability to wage war.

At the heart of the exercise is a facility known as combined air operations center (CAOC). The role of the CAOC is so vital that General John Jumper, the air force chief of staff, designated it as an actual weapon system, comparable to that of a bomber or fighter aircraft. Warfighters plan, coordinate, and direct the air campaign translating national objectives into the air tasking order. The CAOC communicates the plan in great detail, from launch times and refueling routes to types of weapons used. During the campaign, time-sensitive targeting decisions are made using real-time technology and decision-chain-management software. The success of the air

Right: Every division of the U.S. Armed Forces participated in Joint Red Flag 2005. Claimed to be the largest U.S. Air Force–sponsored event of the 2005 fiscal year, forces including the U.S. Navy, Army, and Marines fought live and virtual opponents. Aircraft like this EA-6B Prowler were tasked with protecting surface units and other aircraft by jamming hostile radar and communications. The decision to retire the air force's EF-111A Raven and to assign all Department of Defense radar-jamming missions to the Prowler adds to the significance of the EA-6B in joint warfare.

567

E-8C Joint STARS

The E-8C joint surveillance target attack radar system (Joint STARS) is an airborne battle management, command-and-control, intelligence, surveillance, and reconnaissance platform. Its primary mission is to provide theater ground and air commanders with ground surveillance to support attack operations and targeting that contributes to the delay, disruption, and destruction of enemy forces.

Joint STARS evolved from army and air force programs to develop, detect, locate, and attack enemy armor at ranges beyond the forward area of troops. The first two developmental aircraft deployed in 1991 to Operation Desert Storm and also supported Operation Joint Endeavor in December 1995.

Joint STARS supported NATO troops over Bosnia-Herzegovina in 1996, Operation Allied Force from February to June 1999, and Operation Enduring Freedom and Operation Iraqi Freedom in 2003.

The 116th Air Control Wing is America's first "total force" wing, blending National Guard and active-duty airmen into a single unit. The 116th ACW is the only unit that operates the E-8C and the Joint STARS mission. The 17th and final E-8C aircraft was delivered on March 23, 2005.

Using a modified Boeing 707 airframe, the E-8C Joint STARS (joint surveillance target attack radar system) contains a phased-array radar antenna in a twenty-six-foot canoe-shaped radome under the forward part of the fuselage. The aircraft is capable of providing targeting and battle management data to all joint STARS operators in both aircraft and U.S. Army mobile ground station modules (GSMs). Wide-area surveillance and moving target (WAS/MTI) is designed to detect, locate, and identify slow-moving targets and is so precise that it is capable of differentiating wheeled vehicles from track vehicles. This information can assist battle planners in determining the optimum course of action. *USAF*

campaign is monitored with intelligence systems, and assessments flow from the battlefield to military leaders to help shape the next round of plans. During Joint Red Flag, more than 300 personnel per shift (including joint and coalition participants) worked twenty-four hours a day for three weeks to complete the simulated campaign from the CAOC.

Probably one of the most unique aspects of the exercise is the combination of virtual and constructive training. Virtual training is when aircrews participate in an exercise from different locations via linked simulators—they could be located anywhere and "takeoff," "strike targets," and "land" just as if they were flying in an actual aircraft. The mission takes place over the same

Although Red Flag is generally an air force exercise, it is not uncommon for other U.S. forces to participate, thereby offering maximum realism. The U.S. Navy EA-6B Prowler offers radar-jamming capability and has been integrated into Blue Air tactics. Joint Red Flag 2005 not only brought together the navy, but rather joined all aspects of U.S. forces along with the UK, the Netherlands, Canada, and Kuwait for an unprecedented level of realism witnessed by six other nations.

simulated battlefield that members of CAOC are observing (like the Nellis range). Constructive training involves a modeling and simulation system "flying" missions over the area of operations. One person may control many aircraft that show up on monitors at the CAOC as actual sorties. To the exercise participants, each of these missions is "real," as they must vector the aircraft and control them just as if they were actual sorties over the battlefield. The CAOC has operational control over live, virtual, and constructive air assets in order to execute the air tasking order and meet the needs of the joint air component commander.

According to Air Force Chief of Staff General John Jumper, JRF 2005 was the largest distributed mission operations event in history with thirty-four sites and eighteen virtual simulators linked together.

Starting in 2002, the concept of virtual warfare became reality at Joint Red Flag 2005. Known as "Virtual Flag," developers consider this exercise a test of the technology as well as an opportunity to identify lessons for future exercises. Already benefits have included cost savings because operators don't have to move people and machines thousands of miles to conduct training. Far more important are the lives potentially saved by avoiding hazardous conditions as well as having less of an impact on families by not having units deploy for periods of time.

This virtual technology, known as distributed mission operations (DMO), has moved beyond joining a select few pilots utilizing simulators within the same room to large, complex wartime scenarios. Allied participation in the virtual world, as of mid-2005, has been limited to only Canada and the

Master Sergeant Larry Stockton prepares to acknowledge aircraft entering the range should *Challiss*, the awaiting E-3 Sentry AWACS aircraft, have unforeseen communication problems. Besides the typical challenges Blue Air faces against Red Air, working together and providing redundant resources is just one of the many aspects for which the U.S. forces train.

United Kingdom, though further cooperative ventures with other countries are in the works.

As stated by the United States Joint Forces Command (USJFCOM), JRF 2005 has been designed to:

1. Save lives and resources by training our potential deployers in the lessons learned from contemporary military operations so they will arrive in theater as prepared for the conditions as possible;

2. Assess the extent to which a joint force is able to implement the principles outlined in the secretary of defense joint training transformation implementation plan;

3. Develop improved joint training and experimentation capabilities; and

4. Offer recommendations for current doctrine, organization, training, material, leadership, personnel, and facilities (DOTMLPF) in order to more effectively use our existing weapons systems by developing and adapting a new set of doctrine, organizational, and training principles.

Dignitaries from many nations including Australia, Saudi Arabia, Germany, Spain, Italy, and Israel closely watched the Joint Red Flag exercise while countries that actually participated in the program besides the United States included the United Kingdom, the Netherlands, Canada, and Kuwait. Thirty-nine aircraft squadrons including eight U.S. Marine Corps and Navy squadrons and five units from the United Kingdom all flew in JRF 2005. The various participating aircraft included the A/OA-10 Thunderbolt II, B-1B Lancer, B-52 Stratofortress, HC-130H Hercules, E-3 Sentry (AWACS), F-15C Eagle, F-16CJ Viper, KC-135, McDonnell Douglas AV-8B Harrier, Grumman EA-6B Prowler, McDonnell Douglas/Boeing

HH-60G Pave Hawk

The primary mission of the HH-60G Pave Hawk helicopter is to conduct day or night operations into hostile environments to recover downed aircrew or other isolated personnel during war. Because of its versatility, the HH-60G is also tasked to perform "military operations other than war." These tasks include civil search and rescue, emergency aeromedical evacuation, disaster relief, international aid, counterdrug activities, and NASA space shuttle support.

The Pave Hawk is a twin-engine medium-lift helicopter operated by Air Force Special Operations Command, Pacific Air Forces, Air Education and Training Command, Air National Guard, and Air Force Reserve Command.

Pave Hawks have a long history of use in contingencies, starting in Operation Just Cause. During Operation Desert Storm they provided combat search-and-rescue coverage for coalition forces in western Iraq, coastal Kuwait, the Persian Gulf and Saudi Arabia. They also provided emergency evacuation coverage for U.S. Navy SEAL (SEa, Air, and Land) teams penetrating the Kuwaiti coast before the invasion.

During Operation Allied Force, Pave Hawks provided continuous combat search-and-rescue coverage for NATO air forces, and successfully recovered two air force pilots who were isolated behind enemy lines.

In the aircraft's humanitarian relief missions, three Pave Hawks deployed in March 2000 to Mozambique, Africa, to support international flood relief operations. The HH-60s flew 240 missions in seventeen days and delivered more than 160 tons of humanitarian relief supplies.

In October 2003, the continental U.S. search-and-rescue mission transferred to Air Force Special Operations Command. Prior to this, the Pave Hawks were assigned to Air Combat Command.

The versatile Sikorsky H-60 series helicopters have undoubtedly proven themselves indispensable to U.S. forces. The HH-60G Pave Hawk helicopter is designed mainly for combat search-and-rescue (CSAR) along with troop insertion and emergency evacuation. Mixed into various Red Flag missions, the 66th Rescue Squadron (RQS) is routinely tasked with rescuing purposefully downed airmen. The HH-60 flies at speeds of up to 180 mph with an unlimited range due to its aerial refueling capabilities.

F-15E from the 90th Fighter Squadron stationed at Elmendorf AFB, Alaska, blasts off from runway 3L. Encompassing a 66,000-square-mile area, Red Flag—Alaska will provide an immense geographical change for those accustomed to the typical Red Flag exercises held at Nellis AFB. Just as the Nellis Range Complex offers a diversely hostile desert environment, the Alaskan wilderness offers much of the same on the opposite end of the weather scale. Much of the Alaskan range is only accessible by helicopter and the new Red Flag—Alaska makes the most of that by incorporating humanitarian and rescue missions into the syllabus.

F/A-18 Hornet, HH-60 Blackhawk, Panavia GR4 Tornado, and the BAe Nimrod R1.

Of course, not everything takes place over the vast Nevada range. Cannon AFB in New Mexico was host to the British air force, U.S. Army, and over 275 navy personnel, along with ten F/A-18 Hornets and three Northrop Grumman E-2C Hawkeye radar planes, which started arriving at the 27th Fighter Wing for a joint exercise called Roving Sands. This isn't the first time Roving Sands has given pilots the opportunity to practice against the Patriot missile batteries, also giving the army an opportunity to train. The biggest difference is its inclusion into the JRF 2005 exercise and its nightly missions over Fort Bliss, Texas.

With the conclusion of JRF 2005, there are ways to improve the next Joint Red Flag. "There are a number of items we can improve on now that we have completed this huge exercise," said Lieutenant Colonel Murray. "If I have to pick one item to improve on, it would be how we integrate the full training events and the venue-specific training. If we had done this a bit better during the first week of this exercise, we could have been totally successful."

"The smoothing of the flow of information among all participants, U.S. and coalition, is an area we plan to focus on in the future. We did a great job and can look forward to only getting better. This has been a tremendous exercise and has allowed us to take the first of what I hope are many more strides toward fully joint and combined training events."

Just one of dozens of aircraft to launch at night, this F-15 takes off from Nellis Air Force Base on its way to the Nellis Range Complex (NRC) for a few hours of night-time combat. Compared to the Nellis Red Flag Exercise, Red Flag—Alaska will provide five and a half times the amount of airspace, with 66,000 square miles compared to the current 12,000 square miles offered by the NRC.

At the end of JRF 2005, 24,000 sorties were flown, 3,500 to 4,000 were live-combat training missions, 6,000 to 7,000 were flown as virtual sorties, and 18,500 were constructive sorties.

Cope Thunder Becomes Red Flag

Another indication of Red Flag's success is the modification of the already successful Cope Thunder exercise. Just like Red Flag, Cope Thunder was designed to give aircrews their first combat experience by acquainting them with the minimal ten-mission rule. Divided into defensive and offensive forces of red and blue, Cope Thunder involves an additional white force that represents the neutral controlling agency.

Initiated in 1976 at Clark Air Base in the Philippines, Cope Thunder was relocated to Eielson AFB, Alaska when Mount Pinatubo posed a threat to operations in 1992. With the immense airspace over Alaska—almost 66,000 square miles—Eielson, in conjunction with Elmendorf AFB, seemed a logical choice. In addition, Eielson already housed the 353rd Combat Training Squadron (CTS), which controlled and maintained three major military flight-training ranges in Alaska.

The 353rd Combat Training Squadron is responsible for organizing, planning, and executing realistic combat training at Eielson, including Cope Thunder. The 353rd CTS also manages all of the activities on Alaska's three weapons-training ranges, including the incorporation of twenty-eight threat systems and 235 targets for range and exercise operations.The participating Alaskan ranges include the Blair Lakes conventional range, the Yukon tactical and electronic warfare range, and the Oklahoma tactical range. The Blair Lakes conventional range is located approximately twenty-six miles southwest of Eielson AFB and is in isolated subarctic tundra environment that is manned continuously and is normally accessible only by helicopter. The Yukon tactical and electronic warfare range is located fifteen miles east of Eielson and is accessible most of the year. The mountainous range complex is only manned as necessary to provide electronic warfare training. The Oklahoma tactical range is located within the U.S. Army's Cold Region Test Center at Fort Greely and is the largest of the three ranges, encompassing more than five hundred thousand acres of relatively flat, open terrain.

Based on the Boeing 767 airframe, the E-3 Sentry airborne warning and control system (AWACS) is most discernable due to the enormous radome atop the fuselage. With a diameter of thirty feet, the radome enables AWACS to separate maritime and airborne targets from ground and sea clutter returns that limit other present-day radar at a distance of over two hundred miles. First delivered in 1977, sixty-six AWACS aircraft serve with various nations including the United States, United Kingdom, France, Saudi Arabia, and NATO.

Expanding Red Flag–Alaska, humanitarian and supply missions are now incorporated into the syllabus. Assisting with humanitarian efforts, both simulated and in reality, is the newest cargo aircraft in the U.S. Air Force inventory, the C-17A Globemaster III. Capable of short-field takeoffs and landings, the C-17 is the only cargo aircraft flown by stick instead of the more standard yoke. This small addition only hints at the C-17's incredible maneuverability not only in the air, but on the ground.

Similar to Red Flag, each deployment is based on a two-week curriculum, however most participating Cope Thunder units usually arrive a week before the actual exercise. This extra week lets aircrews fly one or two range-orientation flights, make physical and mental preparations, familiarize themselves with local flying restrictions, receive local safety and survival briefings, and work on developing orientation plans.

In 1998, the Pacific Air Forces and organizers of Cope Thunder invited various allied nations to participate in order to create a more realistic training environment. This expanded exercise, involving other Pacific Rim countries became known as Cooperative Cope Thunder. In its first year, participating countries included the United Kingdom, Australia, Singapore, and Japan. Unlike Red Flag, Cooperative Cope Thunder also concentrated on humanitarian efforts as well as airlift and peacekeeping missions in keeping with PACAF's efforts to enhance international cooperation.

During Cooperative Cope Thunder in 2005, participating nations included Australia, Japan, Germany, Malaysia, Singapore, South Korea, Thailand, and the United Kingdom. Representatives from Bangladesh, India, Mongolia, and Sri Lanka also observed the exercise.

On March 23, 2006, the air force chief of staff announced that the Cope Thunder exercise had been renamed and enhanced to provide complementary training on the same level as the current Red Flag. Now known as Red Flag–Alaska, the idea is to

One of four different camouflage paint schemes adopted by the 64th Aggressor Squadron adorn this F-16C Viper. Maintaining a relatively low operating cost and providing extreme maneuverability and targeting tactics, the F-16 is the perfect aircraft to assume the aggressor role of the Red Team. To combat similar aircraft from the Blue Team, the 65th Aggressor Squadron composed of F-15Cs has been reestablished and will be flying similar camouflage schemes.

provide the air force with a common set of exercises in multiple locations that have standardized and equal training capabilities.

"Red Flag–Alaska is an air force–level exercise that will build on and reinforce air force to air force habitual relationships," said General T. Michael Moseley. "Making this exercise a Red Flag expands joint training operations and opportunities to improve interoperability with our allies."

According to Air Force Link, Red Flag–Alaska is designed to provide the finest training possible by ensuring that fighter pilots and aircrews participate in at least ten sorties in a realistic simulated combat environment. This is accomplished on the world's largest range, complete with more than twenty-nine air-defense systems, unmanned (ground) threat emitters, and fourth-generation air-combat maneuvering instrumentation pods on aircraft, all of which tie into the Yukon mission debriefing system to provide feedback to the pilots.

With the Nellis Range Complex being limited to twelve thousand square miles, the sixty-six thousand square miles in the Alaskan air space made perfect sense. "The use of these ranges is key to our fifth-generation fighters—the F-22A and the Joint Strike Fighter," General Moseley said. "The space

MQ-1 Predator

The MQ-1 Predator is a medium-altitude, long-endurance, remotely piloted aircraft. The MQ-1's primary mission is interdiction and conducting armed reconnaissance against critical, perishable targets. When the MQ-1 is not actively pursuing its primary mission, it acts as the Joint Forces Air Component Commander–owned theater asset for reconnaissance, surveillance, and target acquisition in support of the joint forces commander.

The "M" is the Department of Defense designation for multi-role and "Q" means unmanned aircraft system. The "1" refers to the aircraft being the first of a series of purpose-built remotely piloted aircraft systems.

The Predator system was designed in response to a Department of Defense requirement to provide persistent intelligence, surveillance, and reconnaissance information to the warfighter.

In April 1996, the secretary of defense selected the U.S. Air Force as the operating service for the RQ-1 Predator system. A change in designation from "RQ-1" to "MQ-1" occurred in 2002 with the addition of the armed reconnaissance role.

Operational squadrons are the 11th, 15th, and 17th Reconnaissance Squadrons, Indian Springs Air Force Auxiliary Field, Nevada.

Lined up on the tarmac, RQ-1 Predator aircraft await launch from a remote location in the Middle East. The 11th and 15th Reconnaissance Squadrons based at Creech AFB are responsible for training pilots and maintenance crews tasked with operating the Predator UAV. The UAV program under the guidance of the USAF has enabled the Predator to become one of the most versatile UAV platforms currently in service. *USAF*

Major Mike "Dozer" Shower describes his experiences with the F-22 Raptor along with details as to the future of the technologically advanced aircraft. The F-22A Raptor has seen very limited experience with Red Flag exercises simply due to its lack of adversaries. In describing his involvement with the 64th Aggressor Squadron, Shower said, "Red Air hated us! Get killed, go home, get killed, go home, get killed, go home. . . . they got real tired of that and haven't invited us back since."

available and the strides to enhance training operations will make this more world-class than ever before."

Red Flag–Alaska held its first exercise from April 24 through May 5, 2006, involving the assistance of the veteran 64th Aggressor Squadron from Nellis AFB and the 63rd Fighter Squadron from Luke AFB, Arizona, acting as Red Air. Like the mission of Joint Red Flag, Red Flag–Alaska got a running start utilizing the virtual simulation methods learned from the JRF 2005 experiences. More than 1,500 active duty, Reserve and Air National Guard Airmen, eighty-four aircraft, and an army and navy unit trained for the two-week period in the air force's composite force exercise on the Pacific Alaskan Range Complex.

Left: The CJ model of the F-16 Fighting Falcon had been upgraded with a number of features to improve its suppression of enemy air defenses (SEAD) capabilities. Such upgrades included the joint helmet-mounted cueing system (JHMCS), which shows heads-up display data on the helmet visor and allows the pilot to select a target without changing the aircraft's direction. Another significant addition would be the Link 16 multifunctional information distribution system (MIDS), which allows aircraft to share cockpit data and lets pilots merge into one display what all the airplanes are seeing. Finally, the advanced targeting pod known as the Sniper XR is another upgrade incorporated on the aircraft. It has a forward-looking infrared sensor that displays an infrared image of the target for the pilot. The pod helps with precise delivery of laser-guided munitions by using a laser to determine range to a target and to the ground.

Link 16

Providing an increased sense of operational realism, fighting units from the marines, army, and navy establish a new form of data collection. The joint datalink information combat execution joint test and evaluation (JDICE JT&E) team joined Nellis and the USAF Warfare Center to demonstrate the value of the information it can provide to warfighters during Joint Red Flag.

The JDICE JT&E team is a program operated by the Office of the Secretary of Defense and hosted by the USAF Warfare Center. The team develops and tests tactics, techniques, and procedures along with associated Link 16 modifications that increase awareness at the tactical level, supporting the area of responsibility of a joint force commander.

Utilizing radio signals to transmit information, Link 16 is a secure data link that allows for viewing via cockpit or ground command-and-control display terminals. U.S. assets such as the army, marines, and Special Operations Forces tactics, techniques, and procedures (TTPs) have each been evaluated and tested by the JDICE team and are now being evaluated simultaneously during Red Flag before being implemented in the field. Each TTP focuses on deconfliction and targeting processes that filter data and allow operators to transmit actionable data via Link 16 for faster execution at the tactical level.

Throughout Red Flag, JDICE demonstrated how its joint TTP provided timely, accurate, complete, and tactically significant information to Link 16–equipped platforms providing increased situational awareness of the ground battle preventing substantial friendly and allied losses.

This is the first time that an integrated ground and air war battle has been used in this exercise. JDICE Test Director Colonel Billy Gilstrap commented, "JDICE and the Warfare Center are redefining the way we train for war."

Fifth Generation Aircraft and Beyond

"In fights against the Raptor, everybody dies," said General Ronald Keys, head of Air Combat Command for the U.S. Air Force. "We killed thirty-three F-15Cs and didn't suffer a single loss."

"They didn't see us at all," said Lieutenant Colonel Jim Hecker, commander of the 27th Fighter Squadron at Langley AFB, Virginia, after an exercise with eight F-22s in Nevada in November 2005.

Although the F-22 Raptor became operational on December 15, 2005, the aircraft was not invited to Red Flag exercises, especially after the prior month's overly successful engagements. Major Michael "Dozer" Shower commented in a press conference in Florida, "Red Air hated us! Get killed, go home, get killed, go home, get killed, go home. . . . they got real tired of that and haven't invited us back since." Raptors are tentatively scheduled to participate in their first official Red Flag exercise in late 2006, though it's reported that they will be contributors in a more advisory role.

By Lockheed's calculations, the combination of the F-22 and Lockheed Martin F-35 Joint Strike Fighter was five times more effective than legacy aircraft in most scenarios and could accomplish the same number of targets destroyed with 50 to 70 percent fewer aircraft. This places the legacy fighters in a

Able to surpass the speed of sound without engaging afterburners, The Lockheed Martin Aeronautical Systems F-22A Raptor is the first aircraft designed with supercruise technology. The F-22's maneuverability is second to none with its fly-by-wire technology, thrust-vectoring nozzles and two incredibly powerful Pratt & Whitney F119-PW-100 engines. To maintain a low radar signature, the Raptor has four internal weapons bays capable of launching two AIM-9 Sidewinder missiles, six AIM-120C advanced medium-range air-to-air missiles (AMRAAM) or two one-thousand-pound joint direct-attack munitions (JDAM). Just aft and to the right of the pilot resides a hidden M61A2 20mm cannon.

difficult, almost unfair position when tasked against the newer aircraft. Real-world training would be next to impossible.

In the November 2005 issue of *Air Force Magazine*, Major General Stephen M. Goldfein, commander of the Air Warfare Center at Nellis, commented that it is his job to certify equipment for combat, find the best use for that equipment, teach all the premier schools how to use it, and provide a venue for integrated joint training. Modernizing the air force requires buying new weapons, and Red Flag is where Goldfein will eventually put those weapons to the test.

Goldfein said the Joint Red Flag exercise in 2005 was a huge success and signaled the beginning of a new era in the way

RQ-4 Global Hawk

The Global Hawk unmanned aerial vehicle provides air force and joint battlefield commanders near-real-time, high-resolution intelligence, surveillance, and reconnaissance imagery. In the last year, the Global Hawk provided air force and joint warfighting commanders more than fifteen thousand such images to support Operation Enduring Freedom, flying more than fifty missions and one thousand combat hours to date.

Cruising at extremely high altitudes, Global Hawk can survey large geographic areas with pinpoint accuracy, giving military decision makers the most current information about enemy location, resources, and personnel.

Global Hawk currently is undergoing flight testing at the Air Force Flight Test Center at Edwards Air Force Base, California, with more than 1,700 hours and more than 120 successful sorties flown. The Global Hawk Program, Reconnaissance Systems Program Office, Aeronautical Systems Center is located at Wright-Patterson AFB, Ohio, which assumed total program control on October 1, 1998.

On April 20, 2000, a Global Hawk from Eglin AFB, Florida, participated in two exercises that included its first transoceanic flight to Europe, and first mission flown in one theater of operations while under control from another. According to U.S. Joint Forces Command, during twenty-two individual sorties flown during the year-long series of joint deployment exercises, Global Hawk proved its military worth by providing critical intelligence, surveillance, and reconnaissance capabilities to the warfighting community.

To demonstrate interoperability between U.S. and Australian military systems, Global Hawk flew 7,500 miles nonstop across the Pacific to Australia on April 22–23, 2001, setting new world records for UAV endurance. U.S. and Australian Defence Science Technology Organisation officials evaluated UAV performance and future military potential during eleven sorties in the land-sea environment before it flew home to Edwards AFB, six weeks later.

In March 2001, Global Hawk entered the engineering, manufacturing, and development phase of defense acquisition. Global Hawk is currently deployed supporting Operation Enduring Freedom.

The RQ-4 Global Hawk unmanned aerial vehicle (UAV) was thrust into service shortly after the commencement of Operation Enduring Freedom. While many aircraft undergo years of testing, the RQ-4 went into immediate service flying combat sorties and gathering intelligence information vital for the safe advancement of ground troops. With a ceiling of 65,000 feet and a continuous flight time of over forty hours, the RQ-4 has more than proven itself in combat. *USAF*

Seen through the eyes of night-vision goggles (NVG), the RQ-4 Global Hawk unmanned aircraft is the latest in high-altitude surveillance technology. Due to overseas-combat needs combined with a limited supply of aircraft and qualified personnel, the RQ-4 has had very limited involvement with Red Flag exercises. With the development of Joint Red Flag and other exercise advancements, the Global Hawk UAV will soon have an increased profile. *USAF*

war games are run. But despite the value of the exercise, he said, "All training should not be joint training." Air force crews must have the chance to practice and develop their own expertise without other organizations looking over their shoulders. "If we were constantly encumbered by 'joint,' we would never do it right," Goldfein said.

Along with the inclusion of the F-22 Raptor and the F-35 Joint Strike Fighter, future Red Flag exercises will also employ the capabilities of unmanned aerial vehicles such as the RQ-1 Predator and Northrop Grumman RQ-4 Global Hawk. Such UAVs were reportedly used during the Joint Red Flag exercise with great success. Unfortunately, UAVs are currently overtasked with either overseas duties or training new UAV operators.

Taking the mission to space, in 2001 the 527th Space Aggressor Squadron from Schriever AFB, Colorado, made life miserable for downed airmen and those looking to support a rescue. Their role was to play the space-capable adversary and knock out GPS equipment needed for a successful search-and-rescue mission. As with most nations capable of providing a justifiable air and ground threat, chances are they will also possess a credible space threat.

From the Rumsfeld Space Commission Report released January 11, 2001, to the Air Force Chief of Staff's Aerospace Integration Plan initiated in July 1999, space integration is the way of the future in training exercises. With the incorporation of virtual warfare along with space-related scenarios, the realism of Red Flag exercises seems unlimited.

As real-world threats become more sophisticated and advancements in technology become available to those who provide a greater threat, air force exercises such as Red Flag, Joint Red Flag, and Red Flag–Alaska will continue to adapt and provide the most realistic training possible.

Appendix

Aircraft Specifications

A-10/OA-10 THUNDERBOLT II *(see sidebar, page 40.)*

Both the A-10 and the OA-10 are very maneuverable at low speeds and altitude (under 1,000 feet [303.3 meters]), and are highly accurate weapons-delivery platforms. They can loiter near battle areas for extended periods of time. Their wide combat radius and short takeoff and landing capability permits operations in and out of locations near front lines. Using night-vision goggles, A-10/OA-10 pilots can conduct their missions during darkness.

Thunderbolt IIs have night-vision imaging systems (NVIS), goggle-compatible single-seat cockpits, and a large bubble canopy that provides pilots with 360-degree vision. The pilots are protected by titanium armor that also protects parts of the flight-control system. The redundant primary structural sections allow the aircraft to enjoy better survivability during close air support than did previous aircraft.

The aircraft can survive direct hits from armor-piercing and high-explosive projectiles up to 23mm. Their self-sealing fuel cells are protected by internal and external foam. Manual systems back up their redundant hydraulic flight-control systems. This permits pilots to fly and land if hydraulic power is lost.

The Thunderbolt II can be serviced and operated from bases with limited facilities near battle areas. Many of the aircraft's parts are interchangeable left and right, including the engines, main landing gear, and vertical stabilizers.

Avionics equipment includes communications, inertial navigation systems, fire control and weapons delivery systems, target penetration aids, and night-vision goggles. Their weapons delivery systems include heads-up displays that indicate airspeed, altitude, dive angle, navigation information, and weapons-aiming references; a low-altitude safety and targeting enhancement system (LASTE) that provides constantly computed impact-point freefall ordnance delivery; and Pave Penny laser-tracking pods under the fuselage. The aircraft also have armament control panels, and infrared and electronic countermeasures to handle surface-to-air-missile threats. Installation of global positioning systems is currently underway for all aircraft.

The Thunderbolt II's 30mm GAU-8/A Gatling gun can fire 3,900 rounds a minute and can defeat an array of ground targets that includes tanks. Other equipment included is an inertial navigation system, electronic countermeasures, target penetration aids, self-protection systems, and AGM-65 Maverick and AIM-9 Sidewinder missiles.

Specifications

Primary Function: A-10—close air support; OA-10—airborne forward air control
Builder: Fairchild Republic Company
Power Plants: Two General Electric TF34-GE-100 turbofan engines
Thrust: 9,065 pounds each engine
Wingspan: 57 feet, 6 inches (17.42 meters)
Length: 53 feet, 4 inches (16.16 meters)
Height: 14 feet, 8 inches (4.42 meters)
Speed: 420 miles per hour (Mach 0.56)
Maximum Takeoff Weight: 51,000 pounds (22,950 kilograms)
Ceiling: 45,000 feet (13,636 meters)
Range: 800 miles (695 nautical miles)
Armament: One 30mm GAU-8/A seven-barrel Gatling gun; up to 16,000 pounds (7,200 kilograms) of mixed ordnance on eight under-wing and three under-fuselage pylon stations, including 500-pound (225 kilogram) Mk-82 and 2,000-pound (900 kilogram) Mk-84 series low/high drag bombs, incendiary cluster bombs, combined effects munitions, mine dispensing munitions, AGM-65 Maverick missiles, and laser-guided/electro-optically guided bombs; infrared countermeasure flares; electronic countermeasure chaff; jammer pods; 2.75-inch (6.99 centimeter) rockets; illumination flares and AIM-9 Sidewinder missiles.
Crew: One (pilot)
Unit Cost: $9.8 million (fiscal 1998 constant dollars)
Date Deployed: March 1976
Inventory: Active—A-10, 143; OA-10, 70; Reserve—A-10, 46; OA-10, 6; ANG—A-10, 84; OA-10, 18

B-1B LANCER *(see sidebar, page 72.)*

The B-1B's blended wing/body configuration, variable-geometry wings and turbofan afterburning engines combine to provide long-range maneuverability and high speed with enhanced survivability. Forward wing settings are used for takeoff, landings, air refueling and in some high-altitude weapons-employment scenarios. Aft wing sweep settings—the main combat configuration—are typically used during high-subsonic and supersonic flight, enhancing the B-1B's maneuverability in the low- and high-altitude regimes. The B-1B's speed and superior handling characteristics allow it to seamlessly integrate in mixed-force packages. These capabilities, when combined with its substantial payload, excellent radar targeting system, long loiter time, and survivability, make the B-1B a key element of any joint/composite strike force.

The B-1B's offensive avionics system includes high-resolution synthetic-aperture radar capable of tracking, targeting, and engaging moving vehicles as well as self-targeting and terrain-following modes. In addition, an extremely accurate GPS-aided inertial navigation system enables aircrews to autonomously navigate globally (without the aid of ground-based navigation aids) as well as engage targets with a high level of precision. The recent addition of Combat Track II (CTII) radios provides secure connectivity until Link-16 is integrated on the aircraft. In a time-sensitive targeting environment, the aircrew can receive targeting data from the combined air operations center over CT II, then update mission data in the offensive avionics system to strike emerging targets rapidly and efficiently. This capability was effectively demonstrated during operations Enduring Freedom and Iraqi Freedom.

The B-1B's self-protection electronic jamming equipment complements its low-radar cross-section to form an onboard defense system that supports penetration of hostile airspace. The ALQ-161 electronic countermeasures system detects and identifies the full spectrum of adversary threat emitters then applies the appropriate jamming technique either automatically or through operator manual inputs. Chaff and flares are employed against radar and infrared threat systems.

B-1 capabilities are being enhanced through the completion of the conventional mission upgrade program (CMUP), which ads thirty cluster munitions, a global positioning system receiver, an improved weapons interface that allows the inclusion of joint direct-attack munitions guided weapons and advanced secure radios (ARC-210). Survivability is enhanced by the addition of the ALE-50 towed decoy system, which decoys advanced radar-guided surface-to-air and air-to-air missile systems.

The CMUP adds improved avionics computers, which allow the employment of additional advanced guided precision and non-precision weapons:

thirty wind-corrected munitions dispensers, twelve joint standoff weapons, or twenty-four joint air-to-surface standoff missiles (JASSMs). The B-1 will be able to carry and employ any mix of weapons. The B-1 will also be the first platform to carry the extended-range version of the JASSM.

Specifications

Primary Function: Long-range, multi-role, heavy bomber
Builder: Rockwell International/ Boeing
Power Plants: Four General Electric F-101-GE-102 turbofan engines with afterburner
Thrust: 30,000-plus pounds with afterburner, per engine
Wingspan: 137 feet (41.8 meters) extended forward, 79 feet (24.1 meters) swept aft
Length: 146 feet (44.5 meters)
Height: 34 feet (10.4 meters)
Speed: 900-plus mph (Mach 1.2 at sea level)
Maximum Takeoff Weight: 477,000 pounds (216,634 kilograms)
Ceiling: More than 30,000 feet (9,144 meters)
Range: Intercontinental, unrefueled
Armament: 24 GBU-31 GPS-aided JDAMs (both Mk-84 general-purpose bombs and BLU-109 penetrating bombs) or 24 Mk-84 2,000-pound general-purpose bombs; 8 Mk-85 naval mines; 84 Mk-82 500-pound general-purpose bombs; 84 Mk-62 500-pound naval mines; 30 CBU-87, -89, -97 cluster munitions; 30 CBU-103/104/105 WCMD, 24 AGM-158 JASSMs or 12 AGM-154 JSOWs.
Crew: Four (aircraft commander, copilot, and two weapon systems officers)
Unit Cost: $283.1 million (fiscal 1998 constant dollars)
Date Deployed: June 1985
Inventory: Active, 65 (test, 2)

B-2 SPIRIT *(see sidebar, page 78.)*

The B-2's low-observable, or stealth, characteristics give it the unique ability to penetrate an enemy's most sophisticated defenses and threaten its most valued, and heavily defended, targets. Its capability to penetrate air defenses and threaten effective retaliation will provide a strong, effective deterrent and combat force well into the twenty-first century.

The revolutionary blending of low-observable technologies with high aerodynamic efficiency and large payload gives the B-2 important advantages over existing bombers. Its low-observability provides it greater freedom of action at high altitudes, thus increasing its range and a better field of view for the aircraft's sensors. Its unrefueled range is approximately 6,000 nautical miles (9,600 kilometers).

The B-2's low observability is derived from a combination of reduced infrared, acoustic, electromagnetic, visual, and radar signatures. These signatures make it difficult for the sophisticated defensive systems to detect, track, and engage the B-2. Many aspects of the low-observability process remain classified; however, the B-2's composite materials, special coatings, and flying-wing design all contribute to its stealthiness.

The B-2 has a crew of two pilots, a pilot in the left seat and mission commander in the right, compared to the B-1B's crew of four and the B-52's crew of five.

Specifications

Primary Function: Multi-role heavy bomber
Builder: Northrop Grumman Corporation
Power Plants: Four General Electric F-118-GE-100 engines
Thrust: 17,300 pounds each engine
Wingspan: 172 feet (52.12 meters)
Length: 69 feet (20.9 meters)
Height: 17 feet (5.1 meters)
Speed: High subsonic
Takeoff Weight (Typical): 336,500 pounds (152,634 kilograms)
Ceiling: 50,000 feet (15,240 meters)
Range: Intercontinental, unrefueled
Armament: Conventional or nuclear weapons
Payload: 40,000 pounds (18,144 kilograms)
Crew: Two (pilots)
Unit cost: Approximately $1.157 billion (fiscal 1998 constant dollars)
Date Deployed: December 1993
Inventory: Active, 21 (1 test)

B-52 STRATOFORTRESS *(see sidebar, page 11.)*

In a conventional conflict, the B-52 can perform strategic attack, air interdiction, offensive counter-air, and maritime operations. During Operation Desert Storm, B-52s delivered 40 percent of all the weapons dropped by coalition forces. It is highly effective when used for ocean surveillance, and can assist the U.S. Navy in antiship and mine-laying operations. Two B-52s, in two hours, can monitor 140,000 square miles (364,000 square kilometers) of ocean surface.

Pilots wear night-vision goggles (NVGs) to enhance their vision during night operations. Night-vision goggles provide greater safety during night operations by increasing the pilot's ability to visually clear terrain, avoid enemy radar, and see other aircraft in a covert/lights-out environment.

Starting in 1989, ongoing modifications incorporate a global positioning system, heavy-stores adapter beams for carrying two-thousand-pound munitions, and a full array of advance weapons currently under development.

The use of aerial refueling gives the B-52 a range limited only by crew endurance. It has an unrefueled combat range in excess of 8,800 miles (14,080 kilometers).

The aircraft's flexibility was evident in Operation Desert Storm and again during Operation Allied Force. B-52s struck wide-area troop concentrations, fixed installations and bunkers, and decimated the morale of Iraq's Republican Guard. The Gulf War involved the longest strike mission in the history of aerial warfare when B-52s took off from Barksdale Air Force Base, Louisiana, air-launched conventional cruise missiles and returned to Barksdale—a thirty-five-hour nonstop combat mission.

During Operation Allied Force, B-52s opened the conflict with conventional cruise missile attacks and then transitioned to delivering general-purpose bombs and cluster bomb units on Serbian army positions and staging areas.

Specifications

Primary Function: Heavy bomber
Builder: Boeing
Power Plants: Eight Pratt & Whitney TF33-P-3/103 turbofan engines
Thrust: Up to 17,000 pounds each engine
Wingspan: 185 feet (56.4 meters)
Length: 159 feet, 4 inches (48.5 meters)
Height: 40 feet, 8 inches (12.4 meters)
Speed: 650 miles per hour (Mach 0.86)
Maximum Takeoff Weight: 488,000 pounds (219,600 kilograms)
Ceiling: 50,000 feet (15,151.5 meters)
Range: Unrefueled 8,800 miles (7,652 nautical miles)
Armament: Approximately 70,000 pounds (31,500 kilograms) mixed ordnance—bombs, mines, and missiles (modified to carry air-launched cruise missiles)
Crew: Five (aircraft commander, pilot, radar navigator, navigator, and electronic warfare officer)

Unit Cost: $53.4 million (fiscal 1998 constant dollars)
Date Deployed: February 1955
Inventory: Active, 85; Reserve, 9

C-130 HERCULES (see sidebar, page 13.)

Using its aft loading ramp and door, the C-130 can accommodate a wide variety of oversized cargo, including utility helicopters, six-wheeled armored vehicles, standard palletized cargo, and military personnel. In an aerial delivery role, it can airdrop loads up to 42,000 pounds or use its high-flotation landing gear to land and deliver cargo on rough dirt strips.

The flexible design of the Hercules enables it to be configured for many different missions, allowing one aircraft to perform the role of many. Much of the special mission equipment added to the Hercules is removable, allowing the aircraft to revert back to its cargo delivery role if desired.

The C-130J is the latest addition to the C-130 fleet and will replace aging C-130Es. The C-130J incorporates state-of-the-art technology to reduce manpower requirements, lower operating and support costs, and provides life-cycle cost savings over earlier C-130 models. Compared to older C-130s, the J model climbs faster and higher, flies farther at a higher cruise speed, and takes off and lands in a shorter distance. The C-130J-30 is a stretch version, adding fifteen feet to fuselage, increasing usable space in the cargo compartment.

C-130J/J-30: Major system improvements include: advanced two-pilot flight station with fully integrated digital avionics; color multifunctional liquid crystal displays and heads-up displays; state-of-the-art navigation systems with dual inertial navigation system and global positioning system; fully integrated defensive systems; low-power color radar; digital moving-map display; new turboprop engines with six-bladed, all-composite propellers; digital auto pilot; improved fuel, environmental and ice-protection systems; and an enhanced cargo-handling system.

Specifications

Primary Function: Global airlift
Builder: Lockheed Martin
Power Plants: C-130E, four Allison T56-A-7 turboprops, 4,200 horsepower; C-130J, four Rolls-Royce AE 2100D3 turboprops, 4,700 horsepower
Wingspan: 132 feet, 7 inches (39.7 meters)
Length: C-130E/J, 97 feet, 9 inches (29.3 meters); C-130J-30, 112 feet, 9 inches (34.69 meters)
Height: 38 feet, 10 inches (11. 9 meters)
Speed: C-130E, 345 miles per hour/300 ktas (Mach 0.49) at 20,000 feet (6,060 meters); C-130J, 417 miles per hour/362 ktas (Mach 0.59) at 22,000 feet (6,706 meters); C-130J-30, 410 miles per hour/356 ktas (Mach 0.58) at 22,000 feet (6,706 meters)
Maximum Takeoff Weight: C-130E/J, 155,000 pounds (69,750 kilograms); C-130J-30, 164,000 pounds (74,393 kilograms)
Ceiling: C-130E, 19,000 feet (5,846 meters) with 42,000 pounds (19,090 kilograms) payload; C-130J, 28,000 feet (8,615 meters) with 42,000 pounds (19,090 kilograms) payload; C-130J-30, 26,000 feet (8,000 meters) with 44,000 pounds (20,227 kilograms) payload
Maximum Allowable Payload: C-130E, 42,000 pounds (19,090 kilograms); C-130J, 42,000 pounds (19,090 kilograms); C-130J-30, 44,000 (19,958 kilograms) Range at Maximum Normal Payload: C-130E, 1,150 miles (1,000 nautical miles); C-130J, 2,071 miles (1,800 nautical miles); C-130J-30, 1,956 miles (1,700 nautical miles)
Crew: C-130E, five (two pilots, navigator, flight engineer, and loadmaster); C-130J/J-30, three (two pilots and loadmaster)
Unit Cost: C-130E, $11.9 million; C-130J, $48.5 million (fiscal 1998 constant dollars)
Date Deployed: C-130A, Dec 1956; C-130E, Aug 1962; C-130J Feb 1999
Inventory: Active, 186; Air National Guard (ANG), 222; Air Force Reserve (AFR), 106

E-3 SENTRY (see sidebar, page 50.)

The E-3 Sentry is a modified Boeing 707/320 commercial airframe with a rotating radar dome. The dome is held 11 feet (3.33 meters) above the fuselage by two struts. It contains a radar subsystem that permits surveillance from the earth's surface up into the stratosphere, over land or water. The radar has a range of more than 250 miles (375.5 kilometers) for low-flying targets and farther for aerospace vehicles flying at medium to high altitudes. The radar combined with an identification friend or foe subsystem can look down to detect, identify, and track enemy and friendly low-flying aircraft by eliminating ground clutter returns that confuse other radar systems.

Other major subsystems in the E-3 are navigation, communications, and computers (data processing). Consoles display computer-processed data in graphic and tabular format on video screens. Console operators perform surveillance, identification, weapons control, battle management, and communications functions.

The subsystems on the E-3 Sentry can gather and present broad and detailed battlefield information. Data is collected as events occur. This includes position and tracking information on enemy aircraft and ships, and location and status of friendly aircraft and naval vessels. The information can be sent to major command-and-control centers in rear areas or aboard ships. In time of crisis, this data can be forwarded to the president and secretary of defense in the United States. It is a jam-resistant system that has performed missions while experiencing heavy electronic countermeasures.

In support of air-to-ground operations, the Sentry can provide direct information needed for interdiction, reconnaissance, airlift, and close-air support for friendly ground forces. It can also provide information for commanders of air operations to gain and maintain control of the air battle.

As an air defense system, E-3s can detect, identify, and track airborne enemy forces far from the boundaries of the United States or NATO countries. It can direct fighter-interceptor aircraft to these enemy targets. Experience has proven that the E-3 Sentry can respond quickly and effectively to a crisis and support worldwide military deployment operations.

With its mobility as an airborne warning and control system, the Sentry has a greater chance of surviving in warfare than a fixed, ground-based radar system. The E-3 can fly a mission profile for more than eight hours without refueling. Its range and on-station time can be increased through in-flight refueling and the use of an onboard crew rest area.

Specifications

Primary Function: Airborne surveillance, command, control, and communications
Builder: Boeing
Power Plants: Four Pratt & Whitney TF33-PW-100A turbofan engines
Thrust: 21,000 pounds each engine
Wingspan: 130 feet, 10 inches (39.7 meters)
Length: 145 feet, 8 inches (44 meters)
Height: 41 feet, 4 inches (12.5 meters)
Rotodome: 30 feet in diameter (9.1 meters), 6 feet thick (1.8 meters), mounted 11 feet (3.33 meters) above fuselage
Speed: Optimum cruise 360 mph (Mach 0.48)
Maximum Takeoff Weight: 347,000 pounds (156,150 kilograms
Ceiling: Above 29,000 feet (8,788 meters)
Endurance: More than eight hours (unrefueled)
Crew: Four (flight crew), 13–19 specialists (mission crew)
Unit Cost: $270 million (fiscal 1998 constant dollars)

Date Deployed: March 1977
Inventory: Active, 33 (1 test)

E-8C JOINT STARS *(see sidebar, page 100.)*

The E-8C is a modified Boeing 707-300 series commercial airframe extensively remanufactured and modified with the radar, communications, operations, and control subsystems required to perform its operational mission. The most prominent external feature is the canoe-shaped radome under the forward fuselage that houses the side-looking phased-array antenna.

The radar and computer subsystems on the E-8C can gather and display detailed battlefield information on ground forces. The information is relayed in near-real-time to army and marine common ground stations and to other ground command, control, communications, computers, and intelligence nodes.

The antenna can be tilted to either side of the aircraft where it can develop a 120-degree field of view covering nearly 19,305 square miles (50,000 square kilometers) and is capable of detecting targets at more than 250 kilometers (more than 820,000 feet). The radar also has some limited capability to detect helicopters, rotating antennas, and low, slow-moving fixed-wing aircraft.

As a battle management command-and-control asset, the E-8C can support the full spectrum of roles and missions from peacekeeping operations to major theater war.

Specifications

Primary Function: Airborne battle management
Builder: Northrop Grumman
Power Plants: Four Pratt & Whitney TF33-102C turbofan engines
Thrust: 19,200 pounds each engine
Wingspan: 145 feet, 9 inches (44.4 meters)
Length: 152 feet, 11 inches (46.6 meters)
Height: 42 feet, 6 inches (13 meters)
Speed: Optimum orbit speed 390–510 knots (Mach 0.52–0.65)
Maximum Takeoff Weight: 336,000 pounds (152,409 kilograms)
Ceiling: 42,000 feet (12,802 meters)
Range: Nine hours (unrefueled)
Crew: Four (flight crew), fifteen air force and three army specialists
Unit Cost: $244.4 million (fiscal 1998 constant dollars)
Date Deployed: 1996
Inventory: Total Force wing, 17

F-15 EAGLE *(see sidebar, page 34.)*

The Eagle's air superiority is achieved through a mixture of unprecedented maneuverability and acceleration, range, weapons, and avionics. It can penetrate enemy defense and outperform and outfight any current enemy aircraft. The F-15 has electronic systems and weaponry to detect, acquire, track, and attack enemy aircraft while operating in friendly or enemy-controlled airspace. The weapons and flight-control systems are designed so one person can safely and effectively perform air-to-air combat.

A multimission avionics system sets the F-15 apart from other fighter aircraft. It includes a heads-up display, advanced radar, inertial navigation system, flight instruments, ultrahigh-frequency communications, tactical navigation system, and instrument landing system. It also has an internally mounted tactical electronic-warfare system, "identification friend or foe" system, electronic countermeasures set, and a central computer.

The pilot's heads-up display projects on the windscreen all essential flight information gathered by the integrated avionics system. This display, visible in any light condition, provides information necessary to track and destroy an enemy aircraft without having to look down at cockpit instruments.

The F-15's versatile pulse-Doppler radar system can look up at high-flying targets and down at low-flying targets without being confused by ground clutter. It can detect and track aircraft and small high-speed targets at distances beyond visual range down to close range, and at altitudes down to treetop level. The radar feeds target information into the central computer for effective weapons delivery. For close-in dogfights, the radar automatically acquires enemy aircraft, and this information is projected on the heads-up display. The F-15's electronic warfare system provides both threat warning and automatic countermeasures against selected threats.

An automated weapons system enables the pilot to perform aerial combat safely and effectively, using the heads-up display and the avionics and weapons controls located on the engine throttles or control stick. When the pilot changes from one weapon system to another, visual guidance for the required weapon automatically appears on the heads-up display.

The Eagle can be armed with combinations of air-to-air weapons: AIM-7 Sparrow missiles or AIM-120 advanced medium range air-to-air missiles on its lower fuselage corners, AIM-9 Sidewinder or AIM-120 missiles on two pylons under the wings, and an internal 20mm Gatling gun in the right wing root.

Specifications

Primary Function: Tactical fighter
Builder: McDonnell Douglas/Boeing
Power Plants: Two Pratt & Whitney F100-PW-100, 220, or 229 turbofan engines with afterburners
Thrust: (C/D models) 23,450 pounds each engine
Wingspan: 42.8 feet (13 meters)
Length: 63.8 feet (19.44 meters)
Height: 18.5 feet (5.6 meters)
Speed: 1,875 miles per hour (Mach 2.5 plus)
Maximum Takeoff Weight: (C/D models) 68,000 pounds (30,844 kilograms)
Ceiling: 65,000 feet (19,812 meters)
Range: 3,450 miles (3,000 nautical miles) ferry range with conformal fuel tanks and three external fuel tanks
Armament: One internally mounted M-61A1 20mm, six-barrel cannon with 940 rounds of ammunition; four AIM-9 Sidewinder and four AIM-7 Sparrow air-to-air missiles, or eight AIM-120 AMRAAMs, carried externally
Crew: F-15A/C, one (pilot); F-15B/D/E, two (pilot and weapons systems officer)
Unit Cost: A/B models, $27.9 million (fiscal 1998 constant dollars); C/D models, $29.9 million (fiscal 1998 constant dollars)
Date deployed: July 1972
Inventory: Active, 396; ANG, 126

F-15E STRIKE EAGLE

The aircraft uses two crew members, a pilot, and a weapon systems officer (WSO). Previous models of the F-15 are assigned air-to-air roles; the E model is a dual-role fighter. It has the capability to fight its way to a target over long ranges, destroy enemy ground positions and fight its way out.

The aircraft's navigation system uses a laser gyro and a global positioning system to continuously monitor the aircraft's position and provide information to the central computer and other systems, including a digital moving map in both cockpits.

The APG-70 radar system allows aircrews to detect ground targets from long ranges. During the air-to-surface weapon delivery, the pilot is capable of detecting, targeting, and engaging air-to-air targets while the WSO designates the ground target.

The low-altitude navigation and targeting infrared for night (LANTIRN) system allows the aircraft to fly at low altitudes, at night, and in any weather conditions, to attack ground targets with a variety of precision-guided and unguided weapons. The LANTIRN system gives the F-15E unequaled accuracy in weapons delivery day or night and in poor weather, and consists of two pods attached to the exterior of the aircraft.

The navigation pod contains terrain-following radar that allows the pilot to safely fly at a very low altitude following cues displayed on a heads-up display. This system also can be coupled to the aircraft's autopilot to provide hands-off terrain-following capability.

The targeting pod contains a laser designator and a tracking system that mark an enemy for destruction at long ranges. Once tracking has been started, targeting information is automatically handed off to GPS or laser-guided bombs.

One of the most important additions to the F-15E is the combination of the rear cockpit and its new weapons systems officer. On four screens, this officer can display information from the radar, electronic warfare or infrared sensors, monitor aircraft or weapons status and possible threats, select targets, and use an electronic moving map to navigate. Two hand controls are used to select new displays and to refine targeting information. Displays can be moved from one screen to another, chosen from a menu of display options.

In addition to three similar screens in the front seat, the pilot has a transparent glass heads-up display at eye level that displays vital flight and tactical information. The pilot doesn't need to look down into the cockpit, for example, to check weapon status. At night, the screen is even more important because it displays a video picture nearly identical to a daylight view of the world generated by the forward-looking infrared sensor.

Each of the low-drag conformal fuel tanks that hug the F-15E's fuselage can carry 750 gallons of fuel. The tanks hold weapons on short pylons rather than conventional weapon racks, reducing drag and further extending the range of the Strike Eagle.

Specifications

Primary Function: Air-to-ground attack aircraft
Builder: McDonnell Douglas/Boeing
Power Plants: Two Pratt & Whitney F100-PW-220 or 229 turbofan engines with afterburners
Thrust: 25,000–29,000 pounds each engine
Wingspan: 42.8 feet (13 meters)
Length: 63.8 feet (19.44 meters)
Height: 18.5 feet (5.6 meters)
Speed: Mach 2.5-plus
Maximum Takeoff Weight: 81,000 pounds (36,450 kilograms)
Service Ceiling: 50,000 feet (15,000 meters)
Range: 2,400 miles (3,840 kilometers) ferry range with conformal fuel tanks and three external fuel tanks
Armament: One 20mm multibarrel gun mounted internally with 500 rounds of ammunition. Four AIM-7 Sparrow missiles and four AIM-9 Sidewinder missiles, or eight AIM-120 AMRAAM missiles. Any air-to-surface weapon in the air force inventory (nuclear and conventional)
Crew: Two (pilot and weapon systems officer)
Unit cost: $31.1 million (fiscal 1998 constant dollars)
Date deployed: April 1988
Inventory: Active, 217

F-16 FIGHTING FALCON

In an air combat role, the F-16's maneuverability and combat radius exceed that of all potential threat fighter aircraft. It can locate targets in all weather conditions and detect low-flying aircraft in radar ground clutter. In an air-to-surface role, the F-16 can fly more than 500 miles (860 kilometers), deliver its weapons with superior accuracy, defend itself against enemy aircraft, and return to its starting point. All-weather capability allows it to accurately deliver ordnance during non-visual bombing conditions.

In designing the F-16, advanced aerospace science and proven reliable systems from other aircraft such as the F-15 and F-111 were selected. These were combined to simplify the airplane and reduce its size, purchase price, maintenance costs, and weight. The light weight of the fuselage is achieved without reducing its strength. With a full load of internal fuel, the F-16 can withstand up to nine G's, which exceeds the capability of other current fighter aircraft.

The cockpit and its bubble canopy give the pilot unobstructed forward and upward vision, and greatly improved vision over the side and to the rear. The seat-back angle was expanded from the usual 13 degrees to 30 degrees, increasing pilot comfort and gravity-force tolerance. The pilot has excellent flight control of the F-16 through its fly-by-wire system. Electrical wires relay commands, replacing the usual cables and linkage controls. For easy and accurate control of the aircraft during high-G-force combat maneuvers, a side-stick controller is used instead of the conventional center-mounted stick. Hand pressure on the side-stick controller sends electrical signals to actuators of flight control surfaces such as ailerons and the rudder.

Avionics systems include a highly accurate inertial navigation system in which a computer provides steering information to the pilot. The plane has UHF and VHF radios plus an instrument landing system. It also has a warning system and modular countermeasure pods to be used against airborne or surface electronic threats. The fuselage has space for additional avionics systems.

Specifications

Primary Function: Multi-role fighter
Builder: General Dynamics/Lockheed Martin
Power Plant: F-16C/D, one Pratt & Whitney F100-PW-200/220/229 or General Electric F110-GE-100/129
Thrust: F-16C/D, 27,000 pounds
Wingspan: 32 feet, 8 inches (9.8 meters)
Length: 49 feet, 5 inches (14.8 meters)
Height: 16 feet (4.8 meters)
Speed: 1,500 miles per hour (Mach 2 at altitude)
Maximum Takeoff Weight: 37,500 pounds (16,875 kilograms)
Ceiling: Above 50,000 feet (15 kilometers)
Range: More than 2,000 miles ferry range (1,740 nautical miles)
Armament: One M-61A1 20mm multibarrel cannon with 500 rounds; external stations can carry up to six air-to-air missiles, conventional air-to-air and air-to-surface munitions and electronic countermeasure pods.
Crew: F-16C, one (pilot); F-16D, one or two (pilot and weapons systems officer)
Unit cost: F-16A/B, $14.6 million (fiscal 1998 constant dollars); F-16C/D, $18.8 million (fiscal 1998 constant dollars)
Date Deployed: January 1979
Inventory: Active—F-16C/D, 738; Reserve—F-16C/D, 69; ANG—F-16C/D, 473

F-22A RAPTOR *(see sidebar, page 87.)*

A combination of sensor capability, integrated avionics, situational awareness, and weapons provides first-kill opportunity against threats. The F-22A possesses a sophisticated sensor suite allowing the pilot to track, identify, shoot, and kill air-to-air threats before being detected. Significant advances in cockpit design and sensor fusion improve the pilot's situational awareness. In the air-to-air configuration, the Raptor carries six AIM-120 AMRAAMs and two AIM-9 Sidewinders.

The F-22A excels at attacking surface targets. In the air-to-ground configuration the aircraft can carry two one-thousand-pound GBU-32 joint direct-attack munitions internally and will use onboard avionics for navigation and weapons-delivery support. In the future, air-to-ground capability will be enhanced with the addition of an upgraded radar and up to eight small-diameter bombs. The Raptor will also carry two AIM-120s and two AIM-9s in the air-to-ground configuration.

Advances in low-observable technologies provide significantly improved survivability and lethality against air-to-air and surface-to-air threats.

The F-22A engines produce more thrust than any current fighter engine. The combination of sleek aerodynamic design and increased thrust allows the F-22A to cruise at supersonic airspeeds (greater than Mach 1.5) without using an afterburner—a characteristic known as supercruise. Supercruise greatly expands the F-22A's operating envelope in both speed and range over current fighters, which must use a fuel-consuming afterburner to operate at supersonic speeds.

The sophisticated F-22A design, advanced flight controls, thrust vectoring, and high thrust-to-weight ratio provide the capability to outmaneuver all current and projected aircraft. The F-22A design has been extensively tested and refined aerodynamically during the development process.

The F-22A's advanced characteristics ensure the fighter's lethality against all advanced air threats. The combination of stealth, integrated avionics, and supercruise drastically shrinks surface-to-air missile engagement envelopes and minimizes enemy capabilities to track and engage the F-22A. Reduced observability and supercruise accentuate the advantage of surprise in a tactical environment.

The F-22A will have better reliability and maintainability than any fighter aircraft in history. An F-22A squadron will require less than half as much airlift as an F-15 squadron to deploy. Increased F-22A reliability and maintainability pays off in less manpower required to fix the aircraft and the ability to operate more efficiently.

Specifications

Primary Function: Air dominance, multi-role fighter
Builder: Lockheed Martin, Boeing
Power Plants: Two Pratt & Whitney F119-PW-100 turbofan engines with after burners and two-dimensional thrust vectoring nozzles
Thrust: 35,000 pounds each engine
Wingspan: 44 feet, 6 inches (13.6 meters)
Length: 62 feet, 1 inch (18.9 meters)
Height: 16 feet, 8 inches (5.1 meters)
Speed: Mach 2
Empty Weight: 40,000-pound class (approximately 18,000 kilograms)
Ceiling: Above 50,000 feet (approximately 15 kilometers)
Armament: One M-61A2 20mm cannon with 480 rounds; side weapon bays can carry two AIM-9 infrared (heat-seeking) air-to-air missiles and main weapon bays can carry (air-to-air loadout) six AIM-120 radar-guided air-to-air missiles or (air-to-ground loadout) two one-thousand-pound GBU-32 JDAMs and two AIM-120 radar-guided air-to-air missiles.
Crew: One (pilot)

F-117A NIGHTHAWK *(see sidebar, page 55.)*

The unique design of the single-seat F-117A provides exceptional combat capabilities. About the size of an F-15 Eagle, the twin-engine aircraft has quadruple redundant fly-by-wire flight controls. Air refuelable, it supports worldwide commitments and adds to the deterrent strength of U.S. military forces.

The F-117A can employ a variety of weapons and is equipped with sophisticated navigation and attack systems integrated into a digital avionics suite that increases mission effectiveness and reduces pilot workload. Detailed planning for missions into highly defended target areas is accomplished by an automated mission planning system developed specifically to take advantage of the unique capabilities of the F-117A.

Specifications

Primary Function: Fighter and attack
Builder: Lockheed Aeronautical Systems
Power Plants: Two General Electric F404 non-afterburning engines
Wingspan: 43 feet, 4 inches (13.2 meters)
Length: 63 feet, 9 inches (19.4 meters)
Height: 12 feet, 9.5 inches (3.9 meters)
Speed: High subsonic
Weight: 52,500 pounds (23,625 kilograms)
Range: Unlimited with air refueling
Armament: Internal weapons carriage
Crew: One (pilot)
Unit Cost: $45 million
Date Deployed: 1982
Inventory: Active, 55

RQ-4 GLOBAL HAWK *(see sidebar, page 114.)*

Once mission parameters are programmed into Global Hawk, the unmanned aerial vehicle can autonomously taxi, take off, fly, remain on station capturing imagery, return, and land. Ground-based operators monitor UAV health and status, and can change navigation and sensor plans during flight as necessary. The UAV can remain airborne for as long as thirty-five hours.

During a typical mission, the aircraft can fly 1,200 miles to an area of interest and remain on station for twenty-four hours. Its cloud-penetrating synthetic aperture radar/ground moving target indicator and electro-optical and infrared sensors can image an area the size of Illinois (40,000 nautical square miles) in just twenty-four hours. Through satellite and ground systems, the imagery can be relayed in near-real-time to battlefield commanders.

More than half the UAV's components are constructed of lightweight, high-strength composite materials, including its wings, wing fairings, empennage, engine cover, engine intake, and three radomes. Its main fuselage is standard aluminum, semi-monocoque construction.

Specifications

Primary Function: Surveillance and reconnaissance
Builder: Northrop Grumman's Ryan Aeronautical Center
Power Plant: One Rolls-Royce Allison turbofan engine
Wingspan: 116 feet (35.3 meters)
Length: 44 feet (13.4 meters)
Height: 15 feet (4.6 meters)
Speed: up to 340 knots (about 400 miles per hour)
Weight: 25,600 pounds (11,612 kilograms)
Ceiling: up to 65,000 feet (19,812 meters)
Range: 12,000 nautical miles
Crew: unmanned, remotely piloted

HH-60G PAVE HAWK *(see sidebar, page 103.)*

The Pave Hawk is a highly modified version of the army Black Hawk helicopter, featuring an upgraded communications and navigation suite that includes satellite, secure voice, and Have Quick communications and integrated inertial navigation, global positioning, and Doppler navigation systems.

All HH-60Gs have an automatic flight-control system, night-vision goggles, lighting, and a forward-looking infrared system that greatly enhances low-level night operations. Additionally, Pave Hawks have color weather radar and an engine/rotor blade anti-ice system that gives the HH-60G adverse weather capabilities.

Pave Hawk mission equipment includes a retractable in-flight refueling probe, internal auxiliary fuel tanks, two crew-served 7.62mm machine guns, and an 8,000-pound (3,600 kilograms) capacity cargo hook. To improve air transportability and shipboard operations, all HH-60Gs have folding rotor blades.

Pave Hawk combat enhancements include a radar warning receiver, infrared jammer, and a flare/chaff countermeasure dispensing system.

HH-60G rescue equipment includes a hoist capable of lifting a 600-pound load (270 kilograms) from a hover height of 200 feet (60.7 meters), and a personnel locating system that is compatible with the PRC-112 survival radio and provides range and bearing information to a survivor's location.

A limited number of Pave Hawks are equipped with an over-the-horizon tactical data receiver that is capable of receiving near real-time mission update information.

Specifications

Primary Function: Combat search-and-rescue and "military operations other than war" in day, night, or marginal weather conditions.
Builder: United Technologies/Sikorsky Aircraft
Power Plants: Two General Electric T700-GE-700 or T700-GE-701C engines with 1,560 to 1,940 horsepower each engine
Rotor Diameter: 53 feet, 7 inches (14.1 meters)
Length: 64 feet, 8 inches (17.1 meters)
Height: 16 feet, 8 inches (4.4 meters)
Speed: 184 miles per hour (294.4 kph)
Maximum Takeoff Weight: 22,000 pounds (9,900 kilograms)
Range: 445 statute miles; 504 nautical miles (unlimited with air refueling)
Armament: Two 7.62mm machine guns
Crew: Four (two pilots, one flight engineer, and one gunner)
Unit Cost: $9.3 million (fiscal 1998 constant dollars)
Date Deployed: 1982
Inventory: Active, 64; ANG, 18; Reserve, 23

KC-10 EXTENDER *(see sidebar, page 58.)*

The KC-10 can transport up to seventy-five people and nearly 170,000 pounds (76,560 kilograms) of cargo a distance of about 4,400 miles (7,040 kilometers) unrefueled. Combined, the KC-10's six fuel tanks carry more than 356,000 pounds (160,200 kilograms) of fuel—almost twice as much as the KC-135 Stratotanker.

Using either an advanced aerial refueling boom, or a hose and drogue centerline refueling system, the KC-10 can refuel a wide variety of U.S. and allied military aircraft within the same mission. The aircraft is equipped with lighting for night operations.

The KC-10's boom operator controls refueling operations through a digital fly-by-wire system. Sitting in the rear of the aircraft, the operator can see the receiver aircraft through a wide window. During boom refueling operations, fuel is transferred to the receiver at a maximum rate of 1,100 gallons (4,180 liters) per minute; the hose and drogue refueling maximum rate is 470 gallons (1,786 liters) per minute. The automatic load alleviation system and independent disconnect system greatly enhance safety and facilitate air refueling. The KC-10 can be air-refueled by a KC-135 or another KC-10 to increase its delivery range.

The large cargo-loading door can accept most air forces' fighter-unit support equipment. Powered rollers and winches inside the cargo compartment permit moving heavy loads. The cargo compartment can accommodate loads ranging from twenty-seven pallets to a mix of seventeen pallets and seventy-five passengers.

Specifications

Primary Function: Aerial tanker and transport
Builder: McDonnell Douglas/Boeing
Power Plants: Three General Electric CF6-50C2 turbofan engines
Thrust: 52,500 pounds, each engine
Wingspan: 165 feet, 4.5 inches (50 meters)
Length: 181 feet, 7 inches (54.4 meters)
Height: 58 feet, 1 inch (17.4 meters)
Speed: 619 miles per hour (Mach 0.825)
Maximum Takeoff Weight: 590,000 pounds (265,500 kilograms)
Ceiling: 42,000 feet (12,727 meters)
Range: 4,400 miles (3,800 nautical miles) with cargo; 11,500 miles (10,000 nautical miles) without cargo
Maximum Cargo Payload: 170,000 pounds (76,560 kilograms)
Maximum Fuel Load: 356,000 pounds (160,200 kilograms)
Crew: Four (pilot, copilot, flight engineer, and boom operator)
Unit Cost: $88.4 million (fiscal 1998 constant dollars)
Date Deployed: March 1981
Inventory: Active, 59

KC-135 STRATOTANKER *(see sidebar, page 15.)*

Four turbofans mounted under 35-degree swept wings power the KC-135 to takeoffs at gross weights up to 322,500 pounds (146,285 kilograms). Nearly all internal fuel can be pumped through the tanker's flying boom, the KC-135's primary fuel transfer method. A special shuttlecock-shaped drogue, attached to and trailing behind the flying boom, may be used to refuel aircraft fitted with probes. An operator stationed in the rear of the plane controls the boom. A cargo deck above the refueling system can hold a mixed load of passengers and cargo. Depending on fuel-storage configuration, the KC-135 can carry up to 83,000 pounds (37,648 kilograms) of cargo.

During the Vietnam War, KC-135 Stratotankers made the air war different from all previous aerial conflicts. Midair refueling brought far-flung bombing targets within reach. Combat aircraft, no longer limited by fuel supplies, were able to spend more time in target areas.

Specifications

Primary Function: Aerial refueling and airlift
Builder: Boeing
Power Plants: KC-135R/T, four CFM International CFM-56 turbofan engines; KC-135E, four Pratt & Whitney TF33-PW-102 turbofan engines
Thrust: KC-135R, 21,634 pounds each engine; KC-135E, 18,000 pounds each engine
Wingspan: 130 feet, 10 inches (39.88 meters)
Length: 136 feet, 3 inches (41.53 meters)
Height: 41 feet, 8 inches (12.7 meters)
Speed: 530 miles per hour at 30,000 feet (9,144 meters)
Maximum Takeoff Weight: 322,500 pounds (146,285 kilograms)
Ceiling: 50,000 feet (15,240 meters)
Range: 1,500 miles (2,419 kilometers) with 150,000 pounds (68,039 kilograms) of transfer fuel; ferry mission, up to 11,015 miles (17,766 kilometers)
Maximum Transfer Fuel Load: 200,000 pounds (90,719 kilograms)
Maximum Cargo Capability: 83,000 pounds (37,648 kilograms), 37 passengers
Typical Crew: Four (pilot, copilot, navigator, boom operator)
Unit Cost: $39.6 million (fiscal 1998 constant dollars)
Date Deployed: August 1956
Inventory: Active, 235; ANG, 220; AFR, 75

MQ-1 PREDATOR *(see sidebar, page 109.)*

The MQ-1 Predator is a system, not just an aircraft. A fully operational system consists of four aircraft (with sensors), a ground-control station, a Predator primary satellite link, and approximately fifty-five personnel for deployed twenty-four-hour operations.

The basic crew of the Predator is one pilot and two sensor operators. They fly the aircraft from inside the ground-control station via a C-band line-of-sight data link or a Ku-band satellite data link for beyond line-of-sight flight. The aircraft is equipped with a color nose camera (generally used by the pilot for flight control), a daytime variable-aperture TV camera, a variable-aperture infrared camera (for low light/night use), and a synthetic aperture radar (SAR) for looking through smoke, clouds, or haze. The cameras produce full-motion video while the SAR produces still-frame radar images.

The MQ-1 Predator carries the multispectral targeting system with inherent AGM-114 Hellfire missile targeting capability and integrates electro-optical, infrared, laser designator, and laser illuminator into a single sensor package. The aircraft can employ two laser-guided Hellfire antitank missiles.

The system is composed of four major components that can be deployed for worldwide operations. The Predator aircraft can be disassembled and loaded into a "coffin." The ground-control system is transportable in a C-130 (or larger) transport aircraft. The Predator can operate on a 5,000-foot by 75-foot (1,524 meter by 23 meter) hard-surface runway with clear line-of-sight. The ground data terminal antenna provides line-of-sight communications for takeoff and landing. The PPSL provides over-the-horizon communications for the aircraft.

An alternate method of employment, remote split operations, uses a smaller version of the ground controls station (GCS) called the launch and recovery GCS. The LRGCS conducts takeoff and landing operations at the forward deployed location while the continental United States–based GCS conducts the mission via extended communications links.

The aircraft includes an ARC-210 radio, an APX-100 IFF/SIF with Mode 4, an upgraded turbocharged engine and glycol-weeping wet wings for ice mitigation. The latest upgrade includes fuel injection, longer wings, dual alternators, and other improvements.

Specifications

Primary Function: Armed reconnaissance, airborne surveillance, and target acquisition
Builder: General Atomics Aeronautical Systems
Power Plant: One Rotax 914 four-cylinder engine producing 101 horsepower
Wingspan: 48.7 feet (14.8 meters)
Length: 27 feet (8.22 meters)
Height: 6.9 feet (2.1 meters)
Speed: 84 miles per hour (70 knots) cruise, up to 135 miles per hour
Maximum Takeoff Weight: 2,250 pounds (1,020 kilograms)
Ceiling: up to 25,000 feet (7,620 meters)
Range: up to 400 nautical miles (454 miles)
Fuel Capacity: 665 pounds (100 gallons)
Payload: 450 pounds (204 kilograms)
Crew: unmanned, remotely piloted
System Cost: $40 million (fiscal 1997 dollars)
Initial Operational Capability: March 2005
Inventory: Active, 57

RC-135V/W RIVET JOINT *(see sidebar, page 24.)*

This aircraft is an extensively modified Boeing C-135. The Rivet Joint's modifications are primarily related to its onboard sensor suite, which allows the mission crew to detect, identify, and geolocate signals throughout the electromagnetic spectrum. The mission crew can then forward gathered information in a variety of formats to a wide range of consumers via Rivet Joint's extensive communications suite.

The interior seats thirty-four people, including the cockpit crew, electronic warfare officers, intelligence operators, and in-flight maintenance technicians.

The Rivet Joint fleet is currently undergoing significant airframe, navigational, and powerplant upgrades, which include re-engining from the TF33 to the CFM-56 engines used on the KC-135R and upgrade of the flight-deck instrumentation and navigational systems to the aircraft modernization program standard. This includes conversion from analog readouts to a digital glass cockpit configuration.

All Rivet Joint airframe and mission systems modifications are overseen by L-3 Communications (previously Raytheon), under the oversight of the Air Force Materiel Command.

Specifications

Primary Function: Reconnaissance
Builder: L-3 Communications
Power Plants: Four CFM International F108-CF-201 high-bypass turbofan engines
Thrust: 21,600 pounds each engine
Wingspan: 131 feet (39.9 meters)
Length: 135 feet (41.1 meters)
Height: 42 feet (12.8 meters)
Speed: 500-plus miles per hour (Mach 0.66)
Maximum Takeoff Weight: 297,000 pounds (133,633 kilograms)
Unrefueled Range: 3,900 miles (6,500 kilometers)
Crew: Five (three pilots and two navigators)
Mission flight crew: 21–27
Date Deployed: Initial RC-135 conversions from 1964–1968; V/W configurations, 1981
Inventory: Active, 14

Allied Aircraft

GR4 TORNADO *(see sidebar, page 38.)*

Designed from the outset as a low-level supersonic aircraft, the Tornado is capable of carrying a wide range of conventional stores, including the air-launched antiradar missile (ALARM), Paveway II and III laser-guided bombs (LGBs). Future plans include carriage of the new Storm Shadow long-range standoff missile and the Brimstone antiarmor missile system. During the 1991 Gulf War, five Tornadoes were modified to carry the new thermal imaging airborne laser designator (TIALD) pod with great success. Modifications to a number of aircraft were carried out to produce the GR1B variant optimized for maritime strike missions with the Sea Eagle antishipping missile.

The Tornado GR4/4A is a world leader in the field of all-weather day-and-night tactical reconnaissance. The aircraft has no cannons mounted in the forward fuselage. Replacing these are a sideways-looking infrared system and a Linescan infrared surveillance system. The standard Tornado GR1 can also fulfill tactical reconnaissance tasks when equipped with a Vicon camera pod.

The Mk-4A dedicated reconnaissance variant has specialized equipment installed in the airframe. On an external fairing on the front fuselage, a forward-looking infrared sensor is installed, its images being projected on a wide-angle heads-up display. A Hughes Raptor reconnaissance pod can be carried. The new avionics system, which features a MIL-STD-1553B databus, is controlled by the main computer to link the new systems and allow complete integration of an improved defensive-aids suite. It also improves navigation and flight performance.

A 1760 weapons bus, which controls the release of a wide range of weapons—such as the Brimstone "smart" antiarmor munition, the Sea Eagle antishipping missile, the ASRAAM air-to-air missile, and the Storm Shadow standoff missile—that provide improved adaptability to future weapons through the missile control and weapons programming units. Other upgrades include a night-vision-goggle-compatible cockpit, a new color multifunction display for the pilot, TIALD (thermal imaging airborne laser designator) for autonomous target acquisition, laser designation facilities, a defense aids subsystem to protect from surface-to-air missiles and radar-directed antiaircraft guns, and GPS. HOTAS (hands-on throttle and stick) may be added in the future. Preliminary work is carried out at the RAF at St. Athan in Wales, and main conversion takes place at BAe Warton, in Lancashire.

Specifications

Country: Germany, Italy, United Kingdom
Primary Function: Multi-role fighter
Builder: Panavia Aircraft GmbH, Germany
Power Plants: Two Turbo-Union RB199-34R turbofan engines
Thrust: 8,700 pounds dry and 14,480 pounds with afterburner
Span: 13.91 meters fully forward, 8.60 meters fully swept
Length: 16.72 meters
Height: 5.95 meters
Maximum Speed: Mach 2.2 at altitude
Maximum Takeoff Weight: Approximately 28,000 kilograms
Radius of Action: 1,390 kilometers (750 nm) with heavy load
Armament: Two internal 27mm Mauser cannons with 180 rounds per gun, plus more than 9,000 kilograms of external stores on seven hardpoints, including: AIM-9 Sidewinder, HARM, ALARM, and AGM-65 Maverick missiles, guided, unguided, and nuclear bombs
Crew: Two (pilot and weapons systems officer)
Customers: The IDS (interdictor-strike) version of Tornado is in service with the Royal Air Force, Luftwaffe, German navy, Italian Air Force, and Royal Saudi Air Force.

British HERCULES C models *(see sidebar, page 20.)*

The C1 is capable of carrying 92 passengers, while the C3, which is 15 feet (4.58 meters) longer than the original C1, can carry 128 passengers or 30 percent more cargo. The maximum payload is 20 tons, or 45,000 pounds, which can be carried over two thousand miles. Following the Falklands War, all RAF Hercules were fitted with an air-to-air refueling probe that can extend their range to over five thousand miles. The freight bay can accommodate a range of wheeled or tracked vehicles or five to seven pallets of general freight. In the aeromedical evacuation role, either sixty-four or eighty-two NATO standard stretchers can be carried. A small number were also fitted out as air-to-air refueling aircraft, but are no longer used by the RAF. The aircraft can operate from unprepared and semi-prepared surfaces by day or night if required.

The C4 is the same size as the older Hercules C3, which features a fuselage 4.58 meters (15 feet) than the original basic aircraft. The Hercules C5 is the new equivalent of the shorter model.

The Hercules C4 and C5s are optimized for economical operation by the introduction of new Allison turboprop engines, six-bladed composite propellers, and a digital engine-control system that increases takeoff thrust by 29 percent and is 15 percent more efficient. Consequently, there is no longer a requirement for the external under-wing fuel tanks to be fitted. An entirely new "glass" cockpit features heads-up displays and four multifunction displays which replace many of the dials and switches of the original aircraft. These displays are compatible with night-vision goggles.

Specifications

Country: United Kingdom
Primary Function: Global airlift
Builder: Lockheed Martin
Power Plants: Four Rolls-Royce Allison turboprop engines
Wingspan: 132 feet, 7 inches (40.41 meters)
Length: C1 and C5, 97 feet, 9 inches (29.79 meters); C3 and C4, 112 feet, 9 inches (34.37 meters)
Top Speed: C1 and C3, 374 miles per hour (602 kph); C4 and C5, 400 miles per hour (640 kph)
Crew: C1 and C3, five or six and up; C4 and C5, three (flight deck crew of two plus one loadmaster)
Accommodation: C1 and C3, up to 92 or 128 troops, 64 paratroops, or 74 stretchers; a maximum payload of up to 43,399 pounds (19,685 kilograms); C4 and C5, up to 128 infantry; 92 paratroops; 8 pallets; or 24 CDS bundles

MiG-29 FULCRUM *(see sidebar, page 75.)*

Similar in size, design, and performance to the American F-16 and McDonnell Douglas F/A-18 fighters, the MiG-29 is in fact superior to its NATO counterparts in some areas and is the world's first aircraft fitted with dual-mode air intakes. While the aircraft is on the ground, the main air intakes are closed, and air is fed through openings on the upper surface of the wing root to prevent ingestion of foreign objects from the runway. The Fulcrum is also the world's first operational front-line fighter to be equipped with turbofan engines, providing a thrust-to-weight ratio greater than one to one for very high maneuverability.

At least as maneuverable as the F-16, with better high-alpha performance, the MiG-29 is also capable of operating from much shorter, more primitive airstrips. Its sophisticated intakes allow it to exceed Mach 2, and it has a faster rate of climb. Its AA-10 Alamo missiles give the MiG-29 a superior BVR (beyond visual range) capability. A helmet-mounted sighting system allows true off-boresight missile-firing capability, freeing the Fulcrum pilot from pointing his nose at the target to lock it up. The integrated fire-control system allows the MiG-29 pilot to detect and track, then lock on to and launch missiles against a target without using radar.

The MiG-29's wings are swept-back and tapered with square tips. Leading edge root extensions (LERX) are wide and curved down to the front. LERX begin on the nose below the mid-mount point, and the wings' trailing edges end at a high-mounted point.

The MiG-29K was initiated in 1984 as a Russian air force development program for a multi-role fighter, and between 1989 and 1991 the MiG-29K underwent tests aboard the Admiral Kuznetsov aircraft-carrying cruiser. The MiG-29K differed from the MiG-29 production model, featuring a new multifunction radar, dubbed Zhuk; a cabin with monochrome display and use of the HOTAS (hands-on throttle and stick) principle; the RVV-AE air-to-air active homing missiles; antiship and antiradar missiles; as well as air-to-ground precision-guided weapons. The MiG-29K program was revived in response to the decision of the Indian navy to acquire the Admiral Gorshkov aircraft carrier. This called for the provision of the ship with a multi-role ship-based arrested-landing fighter of the MiG-29K size. The ship's combat group will include twelve MiG-29K planes. The aircraft has a remote-control system, large-area folding wing, adjustable centerline air intakes with retractable screens protecting the engines during operation from ground airfields, reinforced landing gear, hook, and a corrosion-protected, reinforced fuselage made specifically for deck-based aircraft.

Specifications

Country: Former Soviet Union
Primary Function: All-weather single-seat counter-air fighter
Builder: Moscow Air Production Organization
Power Plants: Two Klimov/Sarkisov RD-33 turbofan engines
Thrust: 22,200 pounds each engine
Wingspan: 36 feet, 5 inches
Length: 56 feet, 10 inches
Height: 15 feet, 6.25 inches
Maximum Speed: Mach 2.3
Weight (empty): 24,030 pounds
Ceiling: 18,400 meters
Cruise range: 905 nautical miles
Range: With 1,500 kilograms of fuel for 255 nautical miles range
Armament: One 30mm GSh-30L cannon with 150 rounds. Six AAMs including a mix of SARH and AA-8 Aphid (R60), AA-10 Alamo (R27T), AA-11 Archer (R73), and FAB 500-M62, FAB-1000, TN-100, ECM Pods, S-24, and AS-12, AS-14
Crew: One (pilot)

NIMROD MR2 *(see sidebar, page 69.)*

The Nimrod was upgraded to the MR2 standard in the early 1980s; while the flight deck and general systems remained the same (apart from the later addition of an air-to-air refueling probe as a result of lessons learned during the Falklands War in 1982), the main underwater and search systems were given a significant upgrade.

The Nimrod has an unrefueled endurance of around ten hours and, although capable of carrying twenty-five people, including two pilots and a flight engineer, two navigators, an air electronics officer (AEO), a sensor team includes seven air electronics operators.

The Nimrod bomb bay carries the Stingray torpedo and the Harpoon missile for the anti-surface unit mission while for search-and-rescue duties the aircraft has a selection of air deliverable multi-seat dinghies, survival packs, and other stores. The aircraft was also fitted to carry Sidewinder air-to-air missiles during the Falklands War (to allow for opportunity attacks on opposing surveillance aircraft more than for self-defense). Internally the aircraft can carry around 150 sonobuoys of several different types.

Although the Nimrod airframe is old, the MR2 remains a potent and respected maritime patrol aircraft; mission system updates will maintain that capability. It has served with distinction in the Falklands War, two Gulf Wars, and in support of the maritime blockade of the Balkans during the Bosnia crisis, while also regularly monitoring Russian naval movements, both subsurface and surface, in the North Atlantic during and since the Cold War. The Nimrod MR2 will continue in service until all squadrons will have been reequipped with updated and re-engined aircraft known as Nimrod MRA4s.

Famous for its role in support of many air-sea rescues, less well known is the secondary role for which a number of aircraft were adapted as the R1. The original maritime equipment was removed from the airframe and replaced with a highly sophisticated and sensitive suite of systems used for reconnaissance and the gathering of electronic intelligence. The ability of the Nimrod to loiter for long periods, following a high-speed dash to the required area of operation, makes the aircraft ideally suited to this task.

Specifications

Country: United Kingdom
Primary Function: Patrol and search and rescue
Builder: Hawker-Siddeley
Power Plants: Four Rolls-Royce Spey turbofan engines
Wingspan: 114 feet, 10 inches (35 meters)
Length: 126 feet, 9 inches (38.63 meters)
Top Speed: 575 mph (926 kph)
Armament: Up to nine torpedoes, bombs, and depth charges internally; Sidewinder missiles can be carried on underwing pylons for self-defense.
Crew: Thirteen

VC10 *(see sidebar, page 30.)*

In the transportation role, the aircraft can accommodate 150 passengers and a crew of nine. The VC10 can be converted easily by use of a large cabin freight door on the forward left side of the aircraft into a passenger/freight or full freight fit. The cabin is capable of holding up to 45,000 pounds (20,500 kilograms) of freight on its permanently strengthened floor, including NATO standard pallets, ground equipment or vehicles. The aircraft also have an aeromedical evacuation capability, where up to seventy-six stretchers may be fitted.

In 1993, the aircraft were converted to the tanker and transport role with the addition of a refueling pod under the outboard section of each wing. The aircraft can carry up to 154,000 pounds (70,000 kilograms) of fuel, utilizing their original eight fuel tanks. The fuel can either be used to feed the aircraft itself or be dispensed to smaller-type fast-jet-type receivers. It is capable of refueling two aircraft at a time from the wing pods. The VC10 C1K can also be refueled from VC10K or Tristar tanker aircraft by use of its air-to-air refueling probe, which is permanently attached to the aircraft nose.

The VC10 C1K is equipped with a modern flight management navigation system and all avionics to allow worldwide operations. The crew comprises two pilots, flight engineer, navigator, and an air load master. Up to three air stewards are carried, depending on the number of passengers on board.

The VC10 K3 and K4 air-to-air refueling fleet is flown by No. 101 Squadron based at RAF Brize Norton. Each aircraft is a three-point tanker, fuel being dispensed from either the two wing hoses or from the single fuselage-mounted refueling point. The wing hoses can transfer fuel at up to 1,000 kilograms per minute and are used to refuel smaller aircraft (such as Tornados or Harriers). The fuselage position can transfer fuel at up to 2,000 kilograms per minute and is usually used to refuel heavy aircraft, although it can also be used by other aircraft.

Each tanker variant of the VC10 carries a different fuel load. The K3 is equipped with fuselage fuel tanks, mounted in what was the passenger compartment, and can carry up to seventy-eight tons; these internal tanks are missing from the K4, which has a maximum fuel load of sixty-eight tons. All the fuel is available to give away to receivers. The aircraft also have a very limited passenger carrying capability. This is used almost exclusively to carry ground crew and other operational support personnel.

Specifications

Country: United Kingdom
Primary Function: Four-engine passenger, freight, or tanker aircraft
Builder: Vickers
Power Plants: Four Rolls-Royce Conway 301 turbofan engines
Wingspan: 146 feet, 2 inches (44.55 meters)
Length: C1K, 158 feet, 8 inches (48.36 meters); K3 and K4, 171 feet, 8 inches (52.32 meters)
Cruise Speed: C1K5, 518 miles per hour (830 kph); K3 and K4, 580 miles per hour (928 kph) at 38,000 feet (11,580 meters)
Crew: C1K, four
Accommodation: C1K, 150 passengers and stewards; up to seventy-six stretchers with six attendants

Index